ADVANCED THERMOSET COMPOSITES

Industrial and Commercial Applications

ADVANCED THERMOSET COMPOSITES

Industrial and Commercial Applications

Edited by

James M. Margolis
Director
Margolis Marketing and Research Company
New York, New York

VAN NOSTRAND REINHOLD COMPANY
─────────────────────── *New York*

SMU LIBRARY

Copyright © 1986 by Van Nostrand Reinhold Company Inc.

Library of Congress Catalog Card Number: 85-3138
ISBN: 0-442-26357-0

All rights reserved. No part of this work covered by the copyright hereon may be reproduced or used in any form or by any means—graphic, electronic, or mechanical, including photocopying, recording, taping, or information storage and retrieval systems—without permission of the publisher

Manufactured in the United States of America

Published by Van Nostrand Reinhold Company Inc.
115 Fifth Avenue
New York, New York 10003

Van Nostrand Reinhold Company Limited
Molly Millars Lane
Wokingham, Berkshire RG11 2PY, England

Van Nostrand Reinhold
480 Latrobe Street
Melbourne, Victoria 3000, Australia

Macmillan of Canada
Division of Gage Publishing Limited
164 Commander Boulevard
Agincourt, Ontario M1S 3C7, Canada

15 14 13 12 11 10 9 8 7 6 4 4 3 2 1

Library of Congress Cataloging in Publishing Data
Main entry under title:

Advanced thermoset composites.

   Includes bibliographies and index.
   1. Thermosetting composites. I. Margolis, James M.
TA418.9.C6284 1985   668.4'22   85-3138
ISBN 0-442-26357-0

*TO MY WIFE
RENA*

PREFACE

Advanced Thermoset Composites: Industrial and Commericial Applications stands out singularly as a book that uniquely addresses state-of-the-art technologies and applications, both current and future. The reader will find a vast source of detailed information on current and potential applications for a variety of reinforced thermoset resin composites.

High technology, now in its infancy, depends largely on materials selections, and advanced thermoset composites are a necessary choice for many high-tech products.

Advanced composites offer outstanding strength levels and even more impressive, high strength-to-weight ratios compared with alternative materials. Corrosion resistance, design freedom, light weight, product versatility, materials availability, and better-than-favorable process economics for many items—all point to a burgeoning growth in the use of these materials.

This book provides the reader with a sensible and useful overview of the advantages of composites, as well as a technical comparison of properties and applications.

The authors, all established experts in their fields, share their knowledge and experience with the reader. The introduction is a source of valuable technical data and insight into new opportunities. The author has solid experience in the entire area of advanced composites.

Fabrication techniques are described by a pioneer and leader who is a world-recognized authority on composite manufacturing. Among his awards is membership in the Plastics Industry Hall of Fame.

Properties are described by an entrepreneur who has combined this spirit with the pragmatism needed for industrial growth. Chapter 3 provides a detailed compendium of practical technical data on composites.

The cutting edge of advanced composites is found in industrial and commercial applications. The essence of this book is contained in the chapters on electrical and electronic applications, automotives, aircraft, process industries, and building construction materials.

Electrical and electronic products represent a lodestar for advanced composite high technology. Written by a team of professionals with combined experience in science, technology, and product applications, the chapter on electrical and electronic applications is a comprehensive report on electronic packaging, composite materials, and processes.

Automotive applications constitute one of the most challenging areas for composites, requiring both a variety of diverse properties and high-volume production economics. The author of Chapter 5 brings his wide knowledge of automotive composites from Detroit where he is a key factor in new materials progress.

Commercial aircraft applications offer one of the most exciting areas of current activity in the advanced composite industry. The author of Chapter 6 is a seasoned veteran on the subject who works for a major commercial airplane manufacturer which is the most advanced company in the world on composites. This chapter covers the uses of advanced composites in commercial airplane construction from early applications to new-generation aircraft and the future.

Process industries and building construction industries are the most popular market areas for advanced composites on a tonnage consumption basis. New resin systems are being introduced to provide more effective chemical corrosion resistance, higher structural strength and dimensional stability such as creep resistance, and competitive (or better-than-competitive) *in situ* costs. The authors of Chapters 7 and 8 are leaders in their fields, with practical understanding of the problems of—and solutions for—advanced composites.

Advanced composites offer exciting possibilities for design engineers. Chapter 9 provides an outlook and perspective on the potential for design engineers who want to be part of the future of the materials revolution. The author is especially capable of viewing the future for designers, since he has been in the field for more than 25 years and is the head of a composite design consulting company.

JAMES M. MARGOLIS

CONTENTS

Preface / vii

1. INTRODUCTION: New Applications from New Materials, *Charles E. Browning* / 1

   Background / 2
   Resins and Fibers / 3
     Matrix Resins / 3
     Epoxies / 3
       Processes / 4 Performance Characteristics / 4 Applications / 5
     Unsaturated Polyesters / 7
       Background / 7 Processes / 8 Performance Characteristics / 9 Applications / 9
     Vinyl Esters / 9
       Background / 9 Processes / 11 Performance Characteristics / 12 Applications / 12
     Polyimides / 13
       Background / 13 Condensation Polyimides / 14 PMR Polyimides / 17 BMI Polyimides / 18
     Reinforcement Fibers / 21
       Fiberglass (Glass)/Background / 21 Fiber Properties / 22 Reinforcement Forms / 23 Composite Properties / 24 Applications / 24
     Boron Fibers / 25
       Background / 25 Fiber Properties / 26 Reinforcement Forms / 26 Composite Properties / 27 Applications / 27
     Carbon/Graphite Fibers / 27
       Background / 27 Fiber Properties / 29 Reinforcement Forms / 29 Composite Properties / 30 Applications / 31
     Aramid Fibers / 31
       Background / 31 Fiber Properties / 32 Reinforcement Forms / 32 Composite Properties / 33 Applications / 33

Miscellaneous Fibers / 34
Fiber FP® / 34 Nextel® / 35 Nicalon® / 35

2. FABRICATION TECHNIQUES: Review and Perspective for Application Engineers, *W. Brandt Goldsworthy* / 38

Introduction / 38
Filament Winding / 39
  Lathe Type / 39
  Whirling Arm / 41
  Racetrack Type / 42
  Tumble (Polar Orbital) Winder / 44
  Ring Winder / 47
  Spherical Winder / 47
Pultrusion / 47
  Off-axis Fiber Orientation / 49
  Materials / 54
  Typical Standard Production Equipment / 56
  Standard Equipment Applications / 56
  Enlarged Capacity Pultruders / 56
  New Directions / 58
  Basic Pultrusion Design Properties / 61
Pulforming / 62
  Straight Pulforming / 62
  Curved Pulforming / 64
Tape Placement / 65
  Six-axis Machine Designed to Wrap Helicopter Blades / 65
  Production Equipment / 68
Summary / 69

3. PROPERTIES AND PERFORMANCE REQUIREMENTS, *James C. Leslie* / 74

Properties of Advanced Composites / 74
Fiberglass Reinforced Composites / 75
  Glass Fibers (Types) / 75
  Available Forms of Glass Reinforcements / 77
  Glass Composite Properties / 79
  Glass Laminate Properties / 79
Kevlar® Reinforced Composites / 83
  Kevlar® Fiber / 83
  Kevlar® 29 Fiber / 84

    Kevlar® 49 Fiber / 84
    Kevlar® 49 Composites / 85
  Carbon/Graphite Reinforced Composites / 94
    Carbon/Graphite Fibers / 95
    Carbon/Graphite Laminate Properties / 98
  Other Reinforcement Fibers / 105
    Silicon Carbide / 107
    Alumina / 107

4. **ELECTRICAL AND ELECTRONIC APPLICATIONS,**
   D. P. Seraphim, D. E. Barr, W. T. Chen, G. P. Schmitt / **110**

   Introduction / 110
   Electronic Packaging Hierarchy / 112
     Two Single-Layer Composites / 113
     Multilayer Composites / 117
     Card-on-Board Package / 120
     Multichip Modules on Planar Package / 120
     Surface-mount Packages / 122
     Flexible Circuits / 126
   Design Factors / 128
     Circuit Lines / 128
     Plated Through Holes / 132
     Programmable Vias / 132
     Power Distribution / 133
   Composite Materials / 133
     Glass Cloth / 134
     Couplers / 134
     Epoxy Resin Systems / 138
     Impregnation and Lamination / 139
     Dimensional and Humidity Effects / 141
     Other Resin Systems / 142
     Interfaces and Adhesion / 144
     Alkaline Oxidation of Copper Surface for Bonding / 145
     Flexible Circuit Materials / 146
     Fire Retardants / 147
     Copper Foils / 149
   Processes / 150
     Process Sequence / 150
     Hole Making / 152
     Laser Hole Making / 154
     Photolithography / 154

Liquid Photoresist / 155
Dr-Film Photoresist / 157
Solder Masks / 161
Screened Coatings / 161
Copper Deposition / 161
Reliability and Controls / 163
Epilogue: The Future for Printed-Circuit Composites / 163

5. AUTOMOTIVE APPLICATIONS OF COMPOSITES, *Peter Beardmore* / **174**

Materials and Design Criteria / 177
Functional Considerations / 182
Methods of Manufacture / 186
Potential of Composites / 191

6. COMMERCIAL AIRCRAFT APPLICATIONS, *John T. Quinlivan and J. Corey McMillan* / **193**

Early Composite Applications / 194
Introduction of Advanced Composites / 198
Advanced Composite Service Evaluations / 201
New-Generation Aircraft / 210
The Future / 221

7. PROCESS INDUSTRY APPLICATIONS, *Joseph S. McDermott* / **227**

Petroleum and Chemical Industries / 228
Storage Tanks / 228
Materials Comparison / 231
Summary / 232
Pressure Vessels and Tubes / 234
Energy and Conservation Industries / 239
Sucker Rods / 240
Injection Piping / 240
Geothermal Power Plants / 241
Temperature-Control Applications / 241
Environmental Protection and Pollution Control Industries / 242
Electrostatic Precipitation / 242
Duct, Scrubber, Absorber, and Stack Liners / 242
Underground Storage Tanks / 243
Agriculture, Papermaking and Mining Industries / 243
Farm Chemical Handling and Implements / 243

CONTENTS xiii

    Food Production/Preservation/Transport Systems / 243
    Livestock Containment / 244
    Paper Mills and Papermaking / 244
    Mining / 244
Conclusion / 247

8. BUILDING CONSTRUCTION MATERIALS, *Andy Green and Tanongsak Besarnsin* **/ 249**

  Applicability of Advanced Fiberglass Reinforced Plastics (FRP) / 250
  Manufacturing Advanced Composites / 251
    Structural Design with Advanced Composites / 252
    Design Considerations and Specifications / 252
    Nature of Loading / 253
    Environmental Conditions / 253
    Expected Life / 253
    Factor of Safety / 253
    Serviceability State / 253
    Design of FRP Structural Components / 253
  Construction Applications / 254
    Cooling Towers / 255
    Walkway Bridges / 257
    Pipes for Cooling Tower Water Distribution System / 258
    Beams / 260
    Panels for Building Cladding / 263
    Decking / 266
    Pre-engineered Buildings / 268
  Outlook / 270

9. OPPORTUNITIES FOR DESIGN ENGINEERS, *Robert E. Coulehan* **/ 272**

  Introduction / 272
  A Case Study / 272
  Electrical Electronic Industry / 274
  Transportation / 275
  Space Structures / 276
  Robotics / 277
  Industrial Processing and Construction / 278
  Conclusion / 279

Index / 281

1
INTRODUCTION
New Applications from New Materials

Charles E. Browning
Materials Laboratory
Air Force Wright Aeronautical Laboratories
Wright-Patterson Air Force Base, Ohio

Composite materials may be defined as materials composed of at least two distinctly dissimilar materials acting in concert. The properties of the composite system are not attainable by the individual components acting alone.

The composite materials of interest herein are those composed of a reinforcing fibrous material and a resin matrix binder. This combination yields a unique class of materials which provide a user (designer, fabricator, etc.) with a unique capability—namely, tailorability—which allows the properties, whether they be performance or processing, to be specifically tailored to the requirement.

Distinctions can be made between the class of composites known as reinforced plastics and the class known as advanced composites. However, for the purposes of this discussion, advanced composites will be considered to include the high modulus fibers, such as graphite and boron, as well as the lower modulus glass fibers combined with a high performance matrix resin such as an epoxy, vinyl ester, etc. It is the combination of these materials that has provided unique properties superior to traditional materials such as aluminum, steel, etc.

Because the properties of composites can be tailored to meet a broad spectrum of desired characteristics, they have found a wide variety of applications in such areas as aerospace, automotive, electronics, industrial, consumer, etc. Many of these applications will be discussed in subsequent chapters.

The major advantage that has driven the widespread use of composites is that of specific properties. When compared to traditional materials, composites are found to have higher moduli per unit weight (specific moduli) and higher strengths per unit weight (specific strengths). These higher specific moduli and strengths can be translated directly into weight savings. These reductions in weight, in turn, result in more efficient structures, reduced energy costs, and reduced material costs.

In addition to these weight savings, composites also offer many other advantages, such as resistance to fatigue and corrosion, low cost fabrication, tailored thermal expansion characteristics and thermal conductivity, damping, and design flexibility by tailoring the reinforcement locations to design requirements.

BACKGROUND

Reinforced plastics experienced a significant growth following World War II. They came to be utilized in the manufacture of boats, cars, furniture, and other consumer and industrial applications. However, their full potential as high performance structural materials was not realized until a technological breakthrough occurred which resulted in the development of a class of reinforced plastics known as advanced composites.

This breakthrough was the result of innovative work conducted at Wright-Patterson Air Force Base, Ohio, in the early 1940s. In this work, a BT-15 composite aft fuselage was designed, fabricated, static tested, and flight tested. This fuselage was a sandwich construction which incorporated glass reinforced polyester face sheets and a balsa wood core. On a specific strength basis, the composite structure was 50% stronger than the baseline aluminum structure. In subsequent work at Wright-Patterson, an AT-6C glass reinforced composite wing was designed, fabricated, and flight tested.

This early pioneering work with glass reinforced composites provided the confidence and impetus for the subsequent work which expanded the advanced composite technology base. This subsequent technology explosion spanned a multitude of disciplines, such as design, processing, materials, testing, etc. An ever increasing usage of advanced composites mirrored this technology explosion.

What may be termed the modern era of advanced composites began about 1960 when boron filaments were developed. In addition to high specific strengths, these reinforcing filaments were capable of providing the high specific stiffnesses missing with glass reinforcement. Further, the introduction of boron filaments was a significant catalyst to advanced composite technology. The technology base that grew out of boron composites demonstrated that these advanced materials had a significant role to play and set the stage for an even further expansion of advanced composite technology.

In terms of a materials technology expansion, the boron filaments were soon followed by the carbon/graphite fibers, aramid fibers, and recent specialty fibers. Concurrent with this expansion of fiber technology was an expansion of resin technology. The initial use of polyester resins was followed by the use of epoxies which eventually became the state-of-the-art, baseline resins for the higher modulus composites. Numerous specialty matrix resins such as high temperature polyimides were also developed for use in advanced composites. Current trends in matrix resin development are addressing the needs for improved toughness and improved environmental durability.

RESINS AND FIBERS

Today's designers and users of advanced composite materials have a wide variety of high performance reinforcing fibers and matrix resins available to them. There are literally limitless opportunities for users of this technology. Matrix resins are available which can satisfy a variety of requirement characteristics, such as sustained high temperature performance, easy processing, toughness, etc. Reinforcement fibers are also available which offer a variety of unique property characteristics. In the following sections, the key property characteristics of the more important matrix resins, reinforcing fibers, and their composites will be discussed.

Matrix Resins

Epoxies. Epoxy resins are thermosetting matrix resins characterized by the epoxide, or oxirane, functionality, $-\overset{O}{\overset{\triangle}{C-C}}-$. With the application of heat, they cure chemically to a cross-linked, insoluble, infusible matrix resin. [Detailed discussion of the chemistry of epoxy resin may be found in reference 1.]

The epoxies are currently the workhorse matrix resins for advanced composites. They are the benchmark against which all new matrix resins are referenced. There are several reasons for this:

1. A wide variety of base resins, curing agents, catalysts, and additives are available [1] to satisfy a variety of property and processing characteristics.
2. Epoxies inherently possess excellent adhesion, mechanical properties, and chemical resistance.
3. Cure shrinkage is relatively low compared to other matrix resins.
4. Epoxies possess excellent processing characteristics:
   - hot-melt prepreggability
   - tacky, drapable prepregs
   - good shelf life
   - addition cures without volatile evolution
   - formulatable for wet filament winding and a variety of unique processes such as resin transfer molding

Processes. There are two main processes for manufacturing epoxy matrix resin based advanced composites. The first process involves preimpregnated epoxy resin composite material forms. The first step in this process is a fiber impregnation process in which the resin impregnates the fibers—individually, as collimated tape, or as a woven form—to yield a prepreg product form. There are a variety of these preimpregnated produce forms available for a user to choose from:

- tows or strands
- monolayer and multilayer tapes
- fabrics
- braids

In the second step of this process, the prepreg form is stacked in layers, wound on a mandrel, braided, or woven to give the desired component geometry. This "lay-up" is then cured, usually under heat and pressure, to give the final product.

The second process is a so-called wet process. In this process, the epoxy resin is combined with the fibers *in situ* to the complete composite fabrication process. This can be via a wet lamination process in which the fibers, in a monolayer sheet form such as woven fabrics or mats, are arranged to a near final shape, impregnated with the wet resin, and then cured. Another type of wet process is a wet winding process in which the fiber tows or strands are drawn through a resin bath, wound onto a mandrel to the required shape, and then cured. Other types of wet processes are molding techniques such as resin transfer molding in which the fibers are shaped in a mold to the final component shape and resin is transferred into the mold to wet the fibers, and then cured.

The final step in all of these composite fabrication processes is the cure. The curing process for epoxy resins consists of the reaction of the epoxy resin with a chemical cross-linking agent (the curing agent or hardener) to yield a cross-linked, three-dimensional polymer network that is insoluble and infusible. It is typically performed in the presence of heat and pressure.

Performance Characteristics. The properties of an epoxy resin are determined by the chemical nature of the reacting molecules, the length or size of the molecules between cross-links (molecular weight between cross-links), and the number of cross-links per unit volume (cross-link density). Because there are such a wide variety of base resins and curing agents, a spectrum of performance characteristics is attainable. Table 1-1 shows properties attainable from one type of epoxy resin—a typical high performance 350°F epoxy resin system used in advanced composites. This system possesses excellent properties, particularly high modulus or stiffness, and retains these properties up to 325°F.

Table 1-1. Typical Properties of a Cured High Performance 350°F Epoxy Resin System Used in Advanced Composites.

Property	Value
Tensile strength, ksi	
RT	12.0
325°F	9.8
Tensile Modulus, Msi	
RT	0.55
325°F	0.34
Strain-to-failure, %	
RT	2.6
325°F	4.2
Glass transition temperature, °C	180
Density, g/cc	1.28

Formulation (Parts by weight)

Epoxy: MY 720 (Ciba-Geigy)—100
Curing agent: diaminodiphenylsulfone—49
Cure: 6 hours at 250°F
   2 hours at 300°F
   2 hours at 350°F

Thermosetting resins such as the epoxies do not melt with heating like thermoplastics, but they do possess an effective upper use temperature at which they suffer losses in stiffness. The upper use temperature is defined by the resin system's glass transition temperature or heat distortion temperature, which, in turn, is determined by the molecular architecture as previously discussed. Table 1-2 shows data for a lower use temperature (+250°F), lower modulus epoxy resin system.

Composite properties attainable with a high performance epoxy resin system and graphite fiber reinforcement are shown in Table 1-3. Composite properties attainable with other epoxy resins and a variety of fiber reinforcements, including glass, boron, other graphites, and aramid fibers, are shown in several tables in the following sections. The epoxies can be seen to be very effective at translating fiber properties to composite properties.

Applications. Because of their outstanding physical properties, epoxy resin based composites find uses in a wide variety of applications which will be dis-

Table 1-2. Typical Properties of a Cured 250°F Epoxy Resin System Used in Advanced Composites.

Property	Value
Tensile strength, ksi	
RT	13.8
250°F	6.7
Tensile Modulus, Msi	
RT	0.41
250°F	0.31
Strain-to-failure, %	
RT	5.5
250°F	9.6
Glass transition temperature, °C	150
Density, g/cc	1.22

Formulation (Parts by weight)

Epoxy: Epon® 828 (Shell)—100
Curing agent: diaminodiphenylsulfone—33
Cure: 6 hours at 250°F
   2 hours at 275°F
   2 hours at 300°F

Table 1-3. Typical Properties Attainable with Epoxy Resin/Graphite Fiber Composites [2].

Property	Value
0° Tensile	
Strength, ksi	278.0
Modulus, Msi	19.7
Strain-to-failure, %	1.4
0° Compression	
Strength, ksi	201.0
Modulus, Msi	19.0
Short beam shear, ksi	17.4
Fiber volume, %	62.0

Material: AS4/3501-6 (Hercules)

cussed in subsequent chapters. Included are such demanding applications as aircraft primary structure, filament wound pressure vessels, etc.

Unsaturated Polyester. *Background.* Unsaturated polyesters, like epoxies, are thermosetting matrix resins which, in the presence of a peroxide catalyst, cure to an insoluble, infusible, cross-linked matrix resin. These resins were discovered in the 1930s and were the matrix resins utilized in the early pioneering work on advanced composites. As noted previously, the early work on aircraft composite structures in the 1940s made use of glass reinforced composites having polyester matrix resins.

A typical unsaturated polyester resin is prepared by the reaction of a glycol (a diol), such as ethylene glycol or propylene glycol, and an anhydride or dibasic acid, such as maleic anhydride, to give an unsaturated structure such as

$$H\left[O-\overset{O}{\underset{\|}{C}}-CH=CH-\overset{O}{\underset{\|}{C}}-O-\overset{CH_3}{\underset{|}{CH}}-CH_2\right]_n O-H$$

This type of resin, in the presence of a peroxide catalyst, will undergo cure via cross-linking reactions at the unsaturated double bonds. This cure is an addition type, with no volatile by-products released, which makes for easier processing.

These polyester resins are most often combined with an unsaturated, reactive monomeric diluent—most frequently, styrene—to give a low viscosity, liquid resin that possesses enhanced processability.

Analogous to the epoxies, the polyesters are very versatile because they can be formulated to satisfy a variety of application requirements. Polyester formulations used in advanced composites typically contain a base polyester resin, reactive diluents, catalyst, accelerator, fillers, and inhibitors. A variety of base resins are available because a variety of acids or anhydrides can be substituted into the polymer backbone. One common type utilizes phthalic anhydride in addition to maleic anhydride and propylene glycol to give the structure:

$$H\left[O-\overset{O}{\underset{\|}{C}}-\underset{}{\bigcirc}-\overset{O}{\underset{\|}{C}}-O-CH_2-\overset{CH_3}{\underset{|}{CH}}-O-\overset{O}{\underset{\|}{C}}-CH=CH-\overset{O}{\underset{\|}{C}}-O-\overset{CH_3}{\underset{|}{CH}}-CH_2\right]_n O-H$$

The chemistry of this polymer can be further tailored to yield specific characteristics as required by a given application. For example, in addition to styrene, there are other popular diluents including vinyl acetate, vinyl toluene, divinyl benzene, triallyl cyanurate, triallyl isocyanurate, and diallylphthalate. Fillers can take on many forms, including clays, minerals, silicas, thermoplastics, etc. These can be used to achieve specific characteristics in the resins, such as flame retardency, shrinkage control, impact resistance, etc.

Peroxides are free radical catalysts used to cure polyester resins. Other catalysts, such as azo compounds, have also been used to cure polyesters, as have radiation and UV energy sources. However, peroxides are by far the most frequently used catalysts. A wide range of peroxide catalysts is available. Frequently used ones include methyl ethyl ketone peroxide (MEKP), benzoyl peroxide (BPO), and *t*-butyl perbenzoate. Quite often, the peroxides are accompanied by accelerators, or promoters which catalyze the breakdown of the catalyst. Popular accelerators include cobalt naphthenate, cobalt octoate, and dimethylaniline.

Inhibitors are essentially free radical scavengers which serve to retard the polymerization. They are used to prevent the resin from aging or reacting before it's desired. Hydroquinone and related compounds are typical inhibitors used in polyester resin formulations.

An optimum catalyst system, which include catalyst, accelerator, and inhibitor, can be chosen to control resin pot life or shelf life, the onset temperature for cure, and cure time.

Processes. There are a variety of processes for manufacturing composites that utilize unsaturated polyester resins. These can be categorized into the general areas of molding, coating, and casting. Many of these are high rate production processes founded on the rapid curing, limited pot life characteristics of the catalyzed polyester resins. However, it is possible to formulate polyester resins to attain a variety of handleability characteristics, including those associated with epoxy based composite prepreg systems [3].

Some of the more important specific processes for fabricating polyester based composites include:

1. Molding or lamination processes in which either preimpregnated fiber material forms are used or the fiber is impregnated *in situ* with the molding. One common sheet form uses a polyester premix coated onto a glass mat to give a sheet molding compound (SMC). These sheets can be stacked in a mold, and shaped and cured under heat and pressure. Molding techniques can also be used in which the reinforcement is stacked in a mold, saturated with resin, and then subjected to consolidation pressure. Another major molding category is press molding, which includes techniques such as injection, compression, and transfer molding.

2. Pultrusion techniques in which the reinforcement is coated with the catalyzed resin and pulled through a heated die to give the cured, shaped part.

3. Filament winding processes similar to those described for the epoxies.
4. Spray techniques in which the reinforcement and catalyzed resin are sprayed onto a mold surface in a buildup technique.
5. Coating techniques in which a force, such as centrifugal, is used to-force the resin/fiberglass material against a tool or mold surface.

Performance Characteristics. Analogous to the epoxies, polyesters can be formulated to yield a broad range of mechanical and physical properties. Table 1-4 shows the range of properties attainable from typical resin system formulations used in high performance composite applications.

Polyesters are most frequently used with glass fiber reinforcement and have also been used with graphite fiber reinforcement [3]. Typical properties attainable with glass reinforced composites are shown in Table 1-5. Properties attainable with graphite fiber reinforced composites are shown in Table 1-6.

Applications. The polyester composites can be seen to offer properties that make them very attractive for may uses. Included are such diverse applications as automotive components, boats, fishing rods, tennis rackets, building products, aircraft components, and appliance housing.

Vinyl Esters. *Background.* Vinyl ester resins are thermosetting polymers which, in the presence of a chemical catalyst or radiation, cure to an insoluble, infusible, cross-linked matrix resin. This is a fairly recent class of resins in that they were intially commercialized in the 1960s. They offer improved chemical resistance, higher heat distortion temperatures, and a better balance of properties than typical unsaturated polyesters.

Table 1-4. Typical Room Temperature Properties of Cured Unsaturated Polyester Resins Used in High Performance Composites [4].

Property	Typical Values
Tensile	
Strength, ksi	6.0–13.0
Modulus, Msi	0.30–0.64
Elongation at break, %	<2
Flexural strength, ksi	8.5–23.0
Compressive strength, ksi	13.0–30.0
Dielectric constant, 60 Hz	3.00–4.36
Specific gravity	1.10–1.46
Water absorption (24 hr, 1/8 in. thick), %	0.15–0.6

Table 1-5. Typical Room Temperature Properties Attainable with Glass Fiber Reinforced Polyester Resin Composites [4].

Property	SMC	Woven Fabric
Tensile		
Strength, ksi	8.0–25.0	30.0–50.0
Modulus, Msi	1.4–2.5	1.5–4.5
Elongation at break, %	3.0	1.0–2.0
Flexural		
Strength, ksi	10.0–36.0	4.00–80.0
Modulus, Msi	1.0–2.2	1.0–3.0
Compressive strength, ksi	15.0–30.0	25.0–50.0
Dielectric constant, 60 Hz	4.4–6.3	4.1–5.5
Specific gravity	1.65–2.60	1.50–2.10
Water absorption (24 hr, 1/8 in. thick specimen), %	0.1–0.25	0.05–0.5

They consist of a chemical structure of the following type:

$$H_2=\overset{R}{\underset{}{C}}-\underset{O}{\overset{\|}{C}}-O-X-O-\underset{O}{\overset{\|}{C}}-\overset{R}{\underset{}{C}}=CH_2$$

where X is the polymer backbone and R is either H (in the case of an acrylate termination) or CH_3 (in the case of a methacrylate termination).

Table 1-6. Typical Room Temperature Properties Attainable with Graphite Fiber Reinforced Polyester Composites [3]

Property	Value
Tensile	
Strength, ksi	220.0
Modulus, Msi	20.2
Elongation, %	1.1
Flexural	
Strength, ksi	272.0
Modulus, Msi	17.7
Interlaminar shear, ksi	13.4
Fiber volume, %	62
Material: DAP polyester/Celion®* 6000	

*®Registered trademark of CCF Inc. of Badische Americas Corporation (BASF).

Compared to the unsaturated polyesters which have unsaturation along the polymer backbone and none at the ends, vinyl esters have only terminal, reactive unsaturation at the ends of the molecule. The vinyl esters also have a relatively lower ester content (usually only at the ends of the molecule) than the thermosetting polyesters which have them all along the polymer molecule.

Predominantly, vinyl esters are derived from epoxy resins by chemically reacting acrylic or methacrylic acid with an epoxy resin (most frequently the diglycidyl ether of bisphenol A type) to yield the typical structure:

$$H_2C=C(R)-C(=O)-O-CH_2-CH(OH)-CH_2-O-C_6H_4-C(CH_3)_2-C_6H_4-O-CH_2-CH(OH)-CH_2-O-C(=O)-C(R)=CH_2$$

where R = −H or −CH$_3$.

The backbone structure and, in turn, the physical and mechanical behavior of the resin can be altered by using different epoxy resins such as novolacs or brominated epoxies. Analogous to the polyesters, vinyl esters are most often combined with unsaturated, reactive diluents, such as styrene, to lower viscosity, improve handleability, and enhance reactivity in the resin.

The vinyl esters are versatile in that they can satisfy a variety of processing and property requirements. This is achieved by the tailorability of the resin. The base resin backbone, the reactive end group, catalyst, accelerator, fillers, and other additives can all be tailored to a specific application. For example, to achieve optimum chemical resistance in the resin, methacrylate end groups are employed.

Vinyl ester resins are cured analogously to unsaturated polyesters. They are polymerized or cross-linked through their unsaturated, reactive vinyl groups by free radicals generated from chemical catalysts, heat, or radiation sources. Cure is a free radical process without evolution of volatile by-products, which makes for easier processing. Popular catalysts used with vinyl esters include methyl ethyl ketone peroxide and benzoyl peroxide. These catalysts are typically used with promoters such as cobalt naphthenate and accelerators such as dimethylaniline.

A variety of radiation sources, including UV and electron beam, have been used to cure vinyl esters. Acrylic esters are preferred for radiation cures because of their greater reactivity. Radiation is preferred over peroxides because of reduced energy requirements, lower shrinkage, faster processing times, ambient temperature cures, and low volatile evolutions.

Processes. Because vinyl esters have many cure characteristics in common with the polyesters, they have many composite fabrication processes in com-

Table 1-7. Typical Room Temperature Properties Attainable with Cured Vinyl Ester Resins [5].

Property	Value	Value
Tensile		
Strength, ksi	12.0	11.7
Modulus, Msi	0.59	0.48
Elongation, %	2	4.5
Flexural		
Strength, ksi	18.4	18.2
Modulus, Msi	0.58	0.5
Heat distortion temperature, °F	300	240
Styrene content, wt %	0	50

mon, including hand lay-up, casting techniques, molding or lamination techniques, and pultrusion. Vinyl esters have also found substantial use in the sheet molding compound (SMC) and bulk molding compound (BMC) processes. Much of the discussion on the cure and processing of unsaturated polyesters is directly applicable to vinyl esters. For example, the choice of the catalyst system can have a major influence on how the materials are processed.

One processing characteristic that is unique to vinyl esters is the frequent use of radiation for curing. This is of particular advantage in coating processes where the benefits of low energy and high rates are achieved.

Performance Characteristics. The vinyl esters can produce a broad range of mechanical and physical properties. Table 1-7 shows typical properties attainable with a high performance vinyl ester resin system.

Typical properties attainable with glass fiber reinforced vinyl ester laminates are shown in Table 1-8.

Composite properties attainable with a typical SMC formulation are shown in Table 1-9.

Applications. The vinyl ester resins can be seen to provide composite properties that make them very attractive for a variety of applications. Of particular note are their high mechanical properties, excellent chemical/corrosion resistance, and elevated temperature resistance. Consequently, vinyl ester based composites are utilized in a variety of applications, including automotive components, appliance housings, pipes, ducts, corrosion resistant linings, storage tanks, and building products.

Table 1-8. Typical Room Temperature Properties of Glass Fiber Reinforced Vinyl Ester Laminates [5].

Property	Value	Value	Value
Tensile			
Strength, ksi	>55.0	28.0	14.0
Modulus, Msi	6.3	1.55	0.9
Elongation, %	0.8	1.4	2.0
Flexural			
Strength, ksi	111.0	32.0	19.0
Modulus, Msi	3.8	1.25	0.7
Glass type	Roving	Woven roving	Mat
Glass content, wt %	75	40	25

Polyimides. *Background.* Polyimides are aromatic-heterocyclic polymeric resins which cure via cross-linking reactions or linear, chain-extension reactions to give high temperature resistant composite matrix resins. The development of this class of elevated temperature resins is due in large part to the activities of the aerospace technical community (government and industry) as it sought improved capability materials that would satisfy the increased demands of high performance aerospace applications.

Polyimides are capable of performance at temperatures far exceeding the best of the materials previously discussed. Whereas the maximum use temperature of epoxies is about 200°C, polyimides can be used at temperatures up to 370°C. The characteristic of polyimides that provides their elevated temperature stability is the aromatic-heterocyclic structure of the polymer backbone:

Table 1-9. Typical Room Temperature Properties Attainable with Vinyl Ester SMC [5].

Property	Value
Glass content, wt %	28
Tensile	
Strength, ksi	13.3
Modulus, Msi	1.58
Flexural	
Strength, ksi	28.0
Modulus, Msi	1.6
Cure time at 302°F, min	1.5

$$\left[-N \underset{\underset{O}{\overset{\|}{C}}}{\overset{\overset{O}{\|}}{\overset{C}{\diagup}}}\hspace{-2pt}R\hspace{-2pt}\underset{\underset{O}{\overset{\|}{C}}}{\overset{\overset{O}{\|}}{\overset{C}{\diagdown}}} N-R'- \right]_n$$

where R and R' can be varied. This type of structure is very thermally and thermooxidatively stable, and provides high glass transition temperatures.

There are three important classes of polyimide matrix resins for advanced composites: (1) condensation polyimides, (2) polymerization of monomeric reactants (PMR) polyimides, and (3) bismaleimides (BMI's). These three will be discussed in the following sections.

Condensation Polyimides. The preparation of a typical condensation polyimide is shown in Figure 1-1. It involves the reaction of an aromatic tetracarboxylic dianhydride (I), such as benzophenone tetracarboxylic dianhydride (BTDA), and an aromatic diamine (II), such as methylene dianiline (MDA). There are two different reaction steps in the polymerization of these materials. In the first reaction, a high molecular weight polyamic acid (III) is formed which is soluble in high boiling point, aprotic solvents such as N-methylpyrrolidone (NMP). This solution, or varnish, is used to impregnate glass or graphite reinforcements to yield prepregs. The next reaction, the cure reaction, involves the cyclization of the amic acid structure to the final imide ring structure (IV). This is accompanied by the release of condensation products such as water and the loss of residual solvent or volatiles. It is this evolution of volatiles during processing which makes the attainment of low void content composites very difficult to achieve with these materials.

The properties of the final product can be tailored by the choice of structures I and II in Figure 1-1. This choice alters the polymer backbone and, in turn, its properties. At the present time, however, there is not a great variety of commercially available monomers from which to choose.

Processes. Typical starting materials for fabricating composites from condensation polyimides are preimpregnated product forms (prepregs) made via processes previously described for epoxies. A polyimide varnish (amic acid in high boiling point solvent) is used to coat, or impregnate, the reinforcement (e.g., glass, graphite, quartz, etc.), giving a prepreg product form that contains residual solvent, as well as volatile condensation products, which the user must remove during composite fabrication. As discussed with the epoxies, the pre-

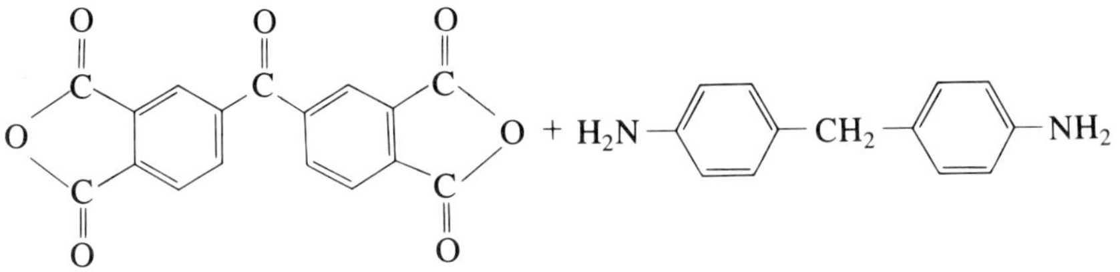

(I) BTDA + (II) MDA

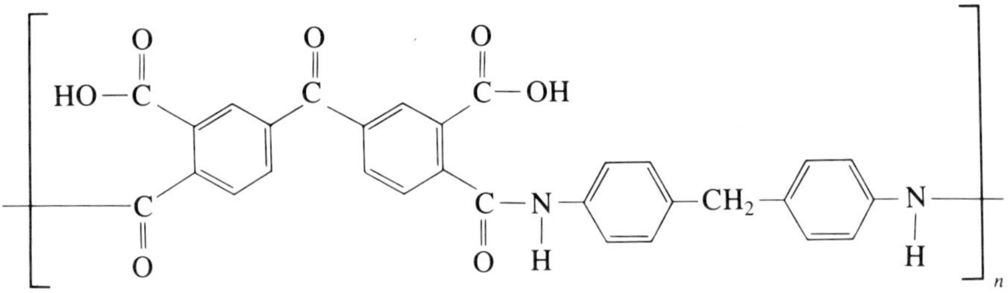

(III) Polyamic Acid

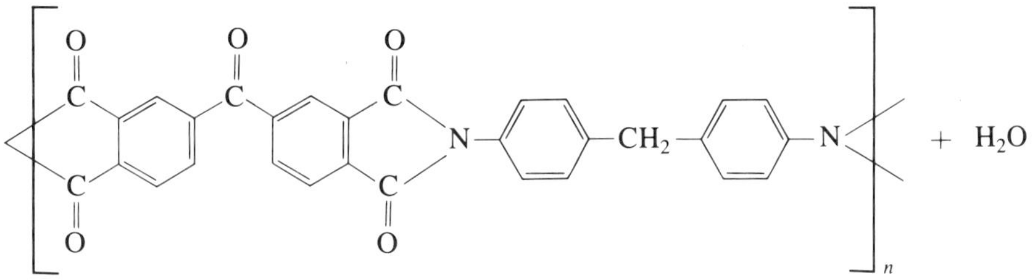

(IV) Polyimide + H₂O

Figure 1-1. Reaction steps for condensation polyimides.

preg is stacked in layers or wound on a mandrel to give the desired component geometry and then cured, under heat and pressure, to give the final product.

The cure of condensation polyimides is much more difficult than that of the epoxies because of the evolution of condensation volatiles and solvent during the cure cycle. In an attempt to minimize porosity in the cured laminate, carefully constructed cure cycles are employed which may contain long hold times at various temperatures to allow for the gradual evolution of volatiles and the application of pressure at precise points in the cycle. Long, stepwise postcure

Table 1-10. Typical Properties Attainable at Room Temperature with Condensation Polyimide/E-Glass Composites [5].

Property	Value
Flexural	
Strength, ksi	85.0
Modulus, Msi	3.12
Tensile strength, ksi	57.0
Elongation, %	1.9
Dielectric constant, 1 MH_z	4.10

cycles are also typically used, with final postcure temperatures determined by the maximum use temperature of the part.

Performance Characteristics. Typical performance characteristics for composites of a condensation polyimide resin reinforced with E-glass fibers are shown in Tables 1-10 and 1-11. It can be seen that these composites possess outstanding properties at room temperature and excellent retention of these properties when exposed to thermooxidative environments.

Applications. Condensation polyimide composites have found use in very demanding applications requiring performance at high temperatures. Many of these applications are aerospace structures, such as radomes and engine components.

Table 1-11. Typical Heat Aging in Air Data for Condensation Polyimide/E-Glass Composites [5].

Condition	Flexural Strength (ksi)	Flexural Modulus (Msi)	Wt Loss (%)
RT	83.0	3.02	—
550°F/2300 hr	41.2	2.63	3.6
550°F/9000 hr	15.0	2.00	12.0
600°F/500 hr	29.0	2.61	2.2
600°F/1850 hr	10.9	2.08	7.9
700°F/100 hr	27.5	1.8	3.0

PMR Polyimides. Work performed at the NASA Lewis Research Center [6] led to the development of a new class of resins known as PMR (polymerization of monomeric reactants) polyimides. These resins contain monomeric reactants which are soluble in low boiling, easily removed solvents such as methanol, resulting in a significant improvement in composite processing over the condensation polyimides.

The components used in PMR-15 resin are shown in Figure 1-2. These monomers are dissolved in a low boiling solvent, such as methanol or ethanol, to give the varnish used to impregnate the reinforcement.

Processes. Starting materials for PMR composites are typically preimpregnated product forms similar to those described in previous sections. The alcohol solution is used to impregnate a variety of reinforcement forms, including graphite tows and cloth, glass rovings and cloth, etc. The prepreg product form is stacked up or filament wound on a mandrel to give the desired geometry and then cured.

The processing of PMR prepregs consists of a series of steps. In the first step, the low boiling solvent (alcohol) is removed at a relatively low tempera-

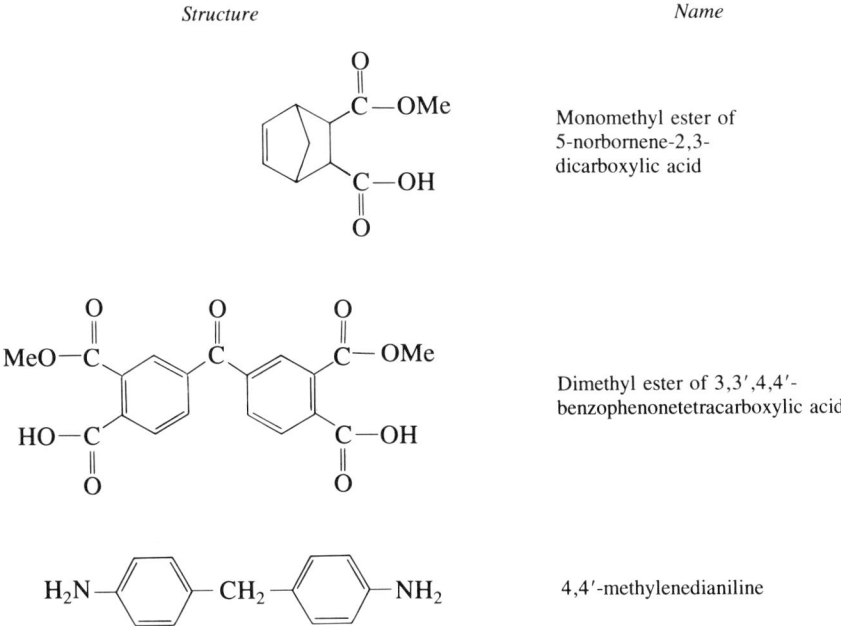

Figure 1-2. Components used in PMR-15 polyimide resin [7].

ture (50–120°C). During cure, the monomers react and imidization initiates at 205°C. This imidized structure is end-capped with the norbornene functionality. At high temperatures (275–350°C), the resin undergoes the final cross-linking reactions by the reverse Diels-Alder reaction of the end-capper, which is an addition reaction without volatile evolution that makes for easier processing. Normally the cure is followed by a postcure treatment. Typically, maximum cure and postcure temperature is 600°F (325°C).

Performance Characteristics. Typical properties of PMR/graphite composites are shown in Table 1-12. Noteworthy is the good retention of properties up to 600°F. Retention of these 600°F properties has been reported for exposure times exceeding 1000 hours at 600°F [5].

Applications. Applications for PMR composites parallel those of the condensation polyimides.

BMI Polyimides. The third important class of polyimides is known as BMI's (bismaleimides). This class was introduced by the French in the late 1960s and is another polyimide that cures via addition reactions without volatile evolution, which results in easier processing.

In this class of polyimide resins, the baseline BMI is formed by the reaction of a diamine (either aromatic or aliphatic) with maleic anhydride:

$$\text{maleic anhydride} + H_2N-R-NH_2 + \text{maleic anhydride} \rightarrow \text{bismaleimide (N-R-N)}$$

R can be varied to achieve a variety of characteristics. This BMI can be used by itself, with other BMI's, and with diamines to form a final resin system. Cure of this final resin can occur via two possible mechanisms depending on the resin composition. If diamines are used, there are two steps in the cure. The first step is a Michael addition reaction of the diamine across the double bond

Table 1-12. Typical Properties of PMR-15/HT-S Composites [8].

Property	Value
Flexural strength, ksi	
RT	214
600°F	129
Flexural modulus, Msi	
RT	16.7
600°F	16.3
Short beam shear, ksi	
RT	15.6
600°F	8.6

$$\text{—R'—NH}_2 + \begin{array}{c}\text{O}\\ \| \\ \text{N—R—} \\ \| \\ \text{O}\end{array}$$

$$\downarrow$$

$$\text{—R'—N}\begin{array}{c}\text{H}\end{array}\begin{array}{c}\text{O}\\ \| \\ \text{N—R—} \\ \| \\ \text{O}\end{array}$$

The second step is the free radical polymerization of the double bonds, which occurs analogously to that described for the unsaturated polyester (see p. 20).

In the case of BMI's without diamines, cure proceeds via the free radical reaction only.

The BMI monomers can be dissolved in solvents to form varnishes for fiber impregnation, or they can be formulated with reactive diluents similarly to the unsaturated polyesters to give solventless, hot-melt resin systems.

Processes. Starting materials for BMI composites can be prepreg product forms similar to those previously described or they can be dry molding compounds. The nature of the prepreg products can vary greatly as a function of resin composition. From dry and boardy to tacky, drapable, epoxy-like prepregs are attainable. All of the product forms are available with a variety of fiber reinforcements (graphite, glass, etc).

$$\text{---R---N}\underset{O}{\overset{O}{\diagdown\diagup}}\xrightarrow{\text{Catalyst}} \text{---R---N}\underset{O}{\overset{O}{\diagdown\diagup}}\cdot$$

$$\text{---R---N}\underset{O}{\overset{O}{\diagdown\diagup}}\cdot + \underset{O}{\overset{O}{\diagdown\diagup}}\text{N---R---}$$

$$\downarrow$$

$$\text{---R---N}\underset{O}{\overset{O}{\diagdown\diagup}}\text{---}\underset{O}{\overset{O}{\diagdown\diagup}}\text{N---R---} \xrightarrow{\text{ETC.}}$$

Cure is a function of the product form. Molding compounds and dry, boardy prepregs are typically cured with high pressure and temperatures of 450–550°F in press cures or injection molding operations. Wet, solvent based prepregs or hot-melt, solventless prepregs are processable with epoxy-like cure conditions in that they can be vacuum-bag, autoclave cured at temperatures of 350°F and pressures of 100 psi. Postcures at temperatures of 425–475°F are typically employed.

Performance Characteristics. Properties attainable with BMI resin/graphite fiber composites are shown in Table 1-13. In general, BMI's are not as thermally stable as condensation or PMR polyimides, but they are much more thermally stable than epoxies. The system shown in the table is a formulated product that possesses epoxy-like processing and epoxy-like room temperature mechanical properties while demonstrating outstanding temperature and moisture resistance. After conditioning in 98% relative humidity for eight weeks, composites tested in flexure at 177°C retained 50% of their room temperature, dry flexural strengths [9].

Applications. BMI composites are utilized in a variety of applications. Their ability to provide higher service temperatures than epoxies while maintaining epoxy-like processing has made them popular for high performance aircraft applications, such as wings or fuselage. BMI composites also see wide usage as printed wiring board materials. Numerous other applications are possible.

Table 1-13. Mechanical and Physical Properties of a BMI/Graphite Composite— V378A/T300 [9].

Property	Value
0° Flexural strength, ksi	
RT	265.0
177°C	197.5
232°C	179.0
288°C	122.0
316°C	107.0
0° Flexural modulus, Msi	
RT	19.8
177°C	20.7
232°C	19.2
288°C	18.4
316°C	17.6
Short beam shear, ksi	
RT	18.3
177°C	10.9
232°C	9.2
288°C	6.4
316°C	5.8
0° RT Tensile	
Strength, ksi	228.9
Modulus, Msi	21.8
Strain, %	1.05
Density, g/cc	1.60
Fiber content, vol %	65

Reinforcement Fibers

Glass. *Background.* As previously discussed, the early pioneering work on structural composites was performed with polyester resins reinforced with glass fibers. These glass fiber reinforced composites were found to exhibit superior properties to traditional materials and, as a result, experienced a phenomenal growth in usage and production. They were the driving force that ushered in the modern era of high performance structural composites. Today, glass is still the reinforcement in largest usage for structural composites.

Two forms of glass fibers are important to the composites industry—the continuous fiber and the discontinuous, or staple, fiber. Both forms start with the same manufacturing process. At the start of the process, molten glass flows into a fiber drawing furnace. Here, it passes through several orifices at the base of

Table 1-14. Composition (wt %) of E-Glass and S-Glass Fibers [5].

Component	E-Glass	S-Glass
Silicon oxide	54.3	64.2
Aluminum oxide	15.2	24.8
Ferrous oxide	—	0.21
Calcium oxide	17.2	0.01
Magnesium oxide	4.7	10.27
Sodium oxide	0.6	0.27
Potassium oxide	—	—
Boron oxide	8.0	0.01
Barium oxide	—	0.2

the drawing furnace to form the individual filaments. In the case of continuous fibers, these extruded filaments are then collected together, quenched, and usually coated with a protective binder known as sizing. These continuous strands are then subjected to subsequent processing steps to form the desired product, e.g., twisted tows, roving, etc. The discontinuous or short staple fibers are produced in a continuous process by passing the continuous strands through a strand chopping device to give short fibers of the desired length.

By varying the orifices in the drawing furnace and the drawing conditions, a variety of filament diameters can be produced. It is, of course, necessary to control these parameters as well as the consistency of the molten glass to maintain a reproducible material.

Chemically, glass is composed predominantly of silica (SiO_2) plus other oxides. Table 1-14 shows the composition for two important glass fiber reinforcements for composites—E-glass and S-glass. E-glass (aluminoborosilicate) compromises the major share of fiberglass production. S-glass (magnesium aluminoborosilicate) is a high performance glass that possesses higher tensile strength, modulus, and use temperature than E-glass. Its development resulted from the demand of aerospace applications for a higher performance fiberglass reinforcement.

Fiber Properties. Glass fibers possess general characteristics which make them attractive for many applications. Their principal features include low cost, excellent electrical properties, good chemical/moisture resistance, good elevated temperature resistance, and very good mechanical properties as shown in Table 1-15.

Table 1-15. Typical Ambient Properties of Glass Fibers [10].

Property	E-Glass	S-Glass
Density, g/cc	2.60	2.49
Tensile		
Strength, ksi	500.0	665.0
Modulus, Msi	10.5	12.6
Elongation, %	4.8	5.4
Coefficient of thermal expansion, 10^{-6} in./in./°F	2.8	3.1
Dielectric constant, 1 MHz	6.33	5.34

Reinforcement Forms. Fiberglass is a very versatile reinforcement in that it is available in a variety of product forms. The more important forms include the following:

- Roving—collections of continuous filaments or strands.
- Yarn—a collection of filaments or strands suitable for use in textile processes. The strands typically have a twist applied to them.
- Woven roving—rovings woven into a fabric.
- Mats—includes mats made from continuous fibers, chopped fibers, or very thin continuous fibers.
- Fabrics—yarns woven together. A substantial variety of fabrics is available depending on the choice of fabric construction, yarn size, number of yarns in the warp direction (lengthwise direction), number of filling yarns (widthwise direction), and weaving pattern (e.g., plain weave, satin weave, etc.)

These forms are available with heat-cleaned fibers or fibers coated with various finishes or coupling agents. These multi-functionality coupling agents react with the filament surface through their silane functionality and react with the matrix resin through another functionality appropriate to the specific resin system. Compared to untreated fibers, these coupling agents provide improved composite mechanical properties, including higher interlaminar shear strengths and resistance to moisture and chemical environments.

All of these reinforcement forms can be combined with a variety of resin systems to yield a variety of product forms. The processing characteristics, including handleability, method of wetting the fibers, cure, etc., are for the most part determined by the specific resin system and were discussed in previous sections.

Table 1-16. Typical Properties of Style 181 S-2 Glass and E-Glass Fabric/Epoxy Resin Composites [10].

Property	S-2 Glass	E-Glass
Flexural		
Strength, ksi	113.6	82.4
Modulus, Msi	4.0	3.5
Compressive strength, ksi	65.7	62.0
Tensile strength, ksi	81.6	57.2
Resin content, wt %	34.4	37.5

Table 1-17. Properties of S-2 Glass and E-Glass Sheet Molding Compounds [10].

Property	S-2 Glass	E-Glass
Flexural		
Strength, ksi	30–34	22–24
Modulus, Msi	1.9–2.2	1.8–2.6
Tensile		
Strength, ksi	18.8	10.8–13.1
Modulus, Msi	2.0–2.2	1.9–2.4
Compressive strength, ksi	32.9–33.3	27.2–36.0
Fiber content, wt %	30	30

Composite Properties. Typical mechanical properties attainable with fiberglass reinforced composites are shown in Tables 1-16 and 1-17. Table 1-16 shows data for epoxy composites reinforced with E-glass and S-2 glass®* (a version of S-glass from Owens-Corning Fiberglass Corporation), while Table 1-17 shows data for a sheet molding compound (SMC) reinforced with the two fibers. It is evident that the higher properties of the S-2 glass translate into higher composite properties.

Applications. Glass composites have found usage in a variety of applications which will be described in subsequent chapters. These include various automotive products and aircraft components, pressure bottles, storage tanks, boating, printed wiring boards, and sporting goods.

---

*Registered trademark of Owens-Corning Fiberglass Corporation.

Boron. *Background.* The introduction of boron filaments was the significant catalyst that ushered in advanced composite technology. Boron filaments possessed the high specific strengths associated with glass, but also possessed very high specific moduli missing with glass fibers. This high specific stiffness characteristic prompted an increased usage of advanced composites plus an expansion of the technology base that resulted in advances in areas such as design, processing, and new materials.

As previously discussed, glass fiber reinforced composites possessed many attributes which made them attractive for replacing traditional materials in many aircraft applications. However, these applications were generally restricted to secondary, noncritical structures because glass composites did not possess the high stiffness necessary for use in primary, flight critical structures such as wings and stabilizers. The development of boron filaments provided a composite material with all of the characteristics, particularly high stiffness, required for use in virtually all primary aircraft structures.

The development of high strength, high modulus boron filaments through a chemical vapor deposition process was initially reported in 1959 by workers at Texaco Experiment Incorporated. Further development work showed that the process could be performed continuously without degrading the fiber properties and that high performance composites could be fabricated with an appropriate matrix resin such as an epoxy. Boron filaments are currently manufactured by Avco Corporation.

Because of the promise offered by the initial developments, the Air Force Materials Laboratory initiated a comprehensive program on boron composite technology. This program included a variety of activities: optimization of the fiber manufacturing process and fiber properties, development of a suitable prepreg product form, development of composite fabrication methodology, generation of an engineering data base, and manufacturing and testing of subscale and full-scale structures. In the mid- to late 1960s, a T-39 center wing box, an F-100 wing cover, and an F-111B wing tip were fabricated from boron composites. These activities resulted in boron composites being established as highly viable engineering composite materials. Even further, the activity established advanced composites as viable engineering materials and established a technology base which facilitated the acceptance of subsequent advanced materials such as graphite fiber composites.

Boron filaments are produced by the chemical vapor deposition of boron on a tungsten substrate in a process that is carried out continuously in a reactor. The tungsten filament (0.5 mil diameter) is resistively heated in the presence of reactive chemical gases—boron trichloride and hydrogen—to give the following reaction:

$$2BCl_3 + 3H_2 \rightleftharpoons 2B + 6HCl$$

The elemental boron produced in the reaction is deposited on the heated tungsten substrate. Three different boron filament diameters—4 mil, 5.6 mil, and 8 mil—have been produced, with the 4 mil being the highest usage fiber. Several reactors can be employed simultaneously depending on the demand. As with other fibers, boron filaments are collected on spools, which facilitates their further use in prepreg processing.

The production of high quality filaments is a function of flaws which can occur during the process. These flaws act as stress risers in the filaments and, in turn, degrade the mechanical properties attainable from the fibers. Control of flaws is necessary to assure consistent, high quality filaments.

The surface morphology of boron filaments has been described as having a corncob appearance. Observable are areas which resembles kernels on an ear of corn. These are believed to be responsible for many of the boron composite properties, including high transverse tensile strength.

Fiber Properties. Typical properties of boron filaments are shown in Table 1-18. In addition to these characteristics, boron filaments are resistant to moisture and corrosive environments, and possess excellent elevated temperature properties.

Reinforcement Forms. Boron filaments are available only in a continuous filamentary form. Because of their inherent stiffness and diameter, they are not amenable to operations such as weaving or braiding. Consequently, the majority of boron filaments are fabricated into composites via prepreg tape technology.

In boron prepreg technology, the boron filaments are converted into continuous boron tapes preimpregnated most frequently with epoxy resins. These prepreg tapes are typically available in widths of $\frac{1}{4}$, 3, 6, and 48 inches, and lengths of up to 1000 feet. The boron filaments are collimated to yield typical fiber contents on the order of 200 per inch of tape width. Prepreg tapes normally have a backing layer of one ply of style 104 glass scrim cloth which improves

Table 1-18. Properties of Boron Filaments [11].

Property	Value
Density, lb/in.^{3}	0.093
Tensile	
Strength, ksi	510.0
Modulus, Msi	58.0
Elongation, %	0.9
Axial coefficient of thermal expansion, 10^{-6} in./in./°F	2.5
Fiber diameter, mils	4.0

*Calculated.

Table 1-19. Properties of Unidirectional Boron/Epoxy Composites [5, 8].

Property	0° Direction	90° Direction
Tensile strength, ksi	270	13
Tensile modulus, Msi	30	3.0
Tensile strain, %	0.7	0.3
Compressive strength, ksi	400	30
Compressive modulus, Msi	30	2.7
Coefficient of thermal expansion (10^{-6} in./in./°F)	2.3	8.0
Density, lb/in.^{3}	0.075	0.075
Fiber content, vol %	60	60

prepreg handling and maintains filament spacing during the lamination and cure process.

Composite Properties. Typical mechanical properties attainable with boron filament reinforced/epoxy resin composites are shown in Table 1-19. Reference 12 contains a very comprehensive set of property data for boron/epoxy composites, including design allowable properties. Of particular note in these composite properties are the high moduli and high compressive strengths that boron/epoxy composites can provide.

Applications. Boron fiber reinforced composites have found applications in a variety of areas. Their high stiffness and compression properties have made them particularly attractive for aircraft applications which have included such demanding structures as wings, and horizontal and vertical stabilizers. Boron fibers have also been used in applications in which they have been combined with other fibers, such as graphite, to give hybrid composites which take advantage of boron's unique properties.

Boron filaments have also found applications in the sporting goods area where they have been used in golf clubs, tennis rackets, squash rackets, and fishing rods.

Carbon/Graphite. *Background.* Carbon/graphite fibers, hereafter referred to as graphite fibers, are high strength, high modulus, lightweight fibers which are the predominant reinforcement in advanced composites today. In general, graphite fibers possess the high specific properties of boron, while offering significant cost and handling advantages over boron. Consequently, they are attractive fibers for use in high performance applications such as primary aircraft structures.

Graphite fibers as a reinforcement for composites had their beginnings in the late 1950s and early 1960s. In those early years, the first high performance fibers were manufactured, and another chapter in the technology of advanced composites was started. As with the boron fibers, the graphite fibers showed such a high potential that a graphite composite technology was initiated which included design, new/improved fibers, weaving, resin matrix materials technology, prepregging, product forms, manufacturing, etc. The growth of this technology is continuing today and will no doubt continue well into the future.

Today's graphite fibers are prepared from either organic fiber precursors or pitch via a pyrolysis-type process. Much of the early work was performed using rayon precursors; however, rayon soon gave way to polyacrylonitrile (PAN) which has become the predominant organic precursor fiber.

Many different processes have been used to convert PAN fibers into graphite fibers. However, the general process consists of four major steps:

1. PAN fiber preparation—special grades of PAN fiber are produced. These fibers are subjected to a stretching operation which orients the fibrillar structure of the PAN and improves its mechanical properties.
2. Stabilization—a process in which the PAN is stabilized against polymer relaxation or softening during subsequent elevated temperature processing steps. The oriented PAN is typically stabilized under tension in an oxidizing atmosphere, resulting in many reactions between the polymer and oxygen.
3. Carbonization—the process by which the stabilized PAN is pyrolyzed into carbon fibers. This process is carried out in an inert atmosphere at temperatures of 1000–1500°C. Tension may be employed to achieve higher degrees of orientation.
4. Graphitization—the highest temperature step, on the order of 2500–3000°C, in which a higher carbon yield and more graphitic microstructure are obtained than in the carbonization step. Tension may be employed in this step to achieve higher degrees of preferred orientation in the fiber. Higher degrees of orientation typically provide higher moduli.

Depending on the specific conditions employed in each step, graphite fibers can be produced having a wide range of properties.

The process of making graphite fibers from a pitch precursor typically consists of the following steps:

1. Conversion of the pitch to a mesophase pitch. Pitch, a high molecular weight, highly aromatic by-product of petroleum distillation processes, is heated in an inert atmosphere for an extended period of time. During this heating, amorphous pitch is transformed into a highly ordered, liquid

crystalline state—mesophase pitch. This orderd state must be achieved to spin highly oriented filaments.
2. Spinning mesophase pitch into fibers—a melt spinning process in which the liquid crystal pitch is extruded through a die to produce fibers.
3. Stabilization—a process similar to that described for PAN. Oxidizing atmospheres or liquids are used.
4. Carbonization—a process similar to that described for PAN.
5. Graphitization—also a process similar to the PAN process.

While the steps 3, 4, and 5 are similar to the PAN process, specific conditions tailored for a pitch process would be used. As with the PAN fibers, conditions can be varied to give a variety of fiber properties.

The high mechanical properties achieved with graphite fibers are attributable to the structure and orientations of the graphite crystals formed during processing. [The graphite crystal consists of planar layers of carbon atoms stacked on top of each other. Within the layers, the carbon atoms are joined by strong covalent bonds. These planar layers are oriented in the direction of the fiber axis as a result of the fiber processing steps.] Varying degrees of orientation result in variations of fiber properties, particularly fiber modulus.

In a final process, graphite fibers are typically subjected to a surface treatment. Most often they are subjected to surface oxidation techniques using liquid oxidizing agents such as nitric acid solutions. These techniques result in increased resin-to-fiber adhesion and, in turn, improved composite properties. Organic coatings or finishes are also frequently used to protect the fiber during operations such as weaving and to improve wetting of the fibers by specific resin systems.

Fiber Properties. As previously noted, a range of properties is available in today's graphite fibers. In fact, in recent years there have been significant improvements in overall mechanical properties, particularly in strain-to-failure. Tables 1-20 and 1-21 show the range of properties available from selected graphite fibers. The intention of the tables is to illustrate representative properties. They are not meant to be all encompassing since a variety of fibers in many forms and having specific properties are available. In addition to those shown in the tables, there are many additional manufacturers of high quality graphite fibers, such as Celanese, Hitco, Hysol-Grafil, Stackpole, Avco, Great Lakes Carbon, FMI, etc. Because of the continued improvements in fiber properties, manufacturer's recent data sheets should be consulted for specific, detailed data.

Reinforcement Forms. Graphite fibers are available in a variety of product forms analogous to those available with glass fibers—continuous, chopped,

Table 1-20. Typical Properties of Carbon/Graphite Fibers [2, 13].

Fiber	Precursor	Tensile Strength (10^3 psi)	Tensile Modulus (10^6 psi)	Strain-To-Failure (%)	Density (lb/in.^{3})
T-300*	PAN	500	33.5	1.50	0.0640
T-700*	PAN	660	36.0	1.80	0.0650
T-50*	PAN	350	57.0	0.70	0.0650
T-55S*	PITCH	250	55.0	0.50	0.0720
P-75S*	PITCH	300	75.0	0.40	0.0720
P-100*	PITCH	325	105.0	0.31	0.0780
P-120*	PITCH	325	120.0	0.27	0.0790
AS4[†]	PAN	578	35.5	1.60	0.0650
AS6[†]	PAN	640	37.9	1.66	0.0655
IM6[†]	PAN	703	44.6	1.66	0.0632
HMS4[†]	PAN	426	52.2	0.86	0.0645

*Union Carbide Corporation.
[†]Hercules Incorporated.

woven fabrics, braids, and mats. The continuous graphite fibers are available as yarns and rovings, and are also available as tows—bundles of numerous filaments. Typical filament counts per tow of 1000, 2000, 3000, 6000, 10,000, 12,000, and greater are available.

All of the fiber forms can be combined with a variety of resin systems to yield a variety of product forms, the most common being a prepreg tape. Prepreg tapes are prepared by collimating tows to the desired width (common widths are 3, 6, 12, and 48 inches) and then impregnating them with a resin to yield a preimpregnated tape—the basic building block of the advanced composite industry.

The various product forms' processing characteristics, including handleability, cure, etc., are for the most part determined by the specific resin system and were discussed in previous sections.

Composite Properties. Typical mechanical properties attainable with graphite composites containing an epoxy matrix resin are shown in Table 1-22 for se-

Table 1-21. Typical Physical Properties of Carbon/Graphite Fibers [13, 14].

Property	PAN	Pitch
Longitudinal coefficient of thermal expansion, 10^{-6} in./in./°F	−0.2 to −0.4	−0.5 to −0.9
Thermal conductivity, Btu ft/hr ft^2 °F	4–40	58–300
Electrical resistivity, ohm-cm × 10^{-4}	9–18	2.5–7.5

Table 1-22. 0° Mechanical Properties Attainable with Unidirectional Graphite/Epoxy (3501-6) Composites [2].

Fiber	Tensile Strength (ksi)	Tensile Modulus (Msi)	Tensile Elongation (%)	Compressive Strength (ksi)	Compressive Modulus (Msi)	Short Beam Shear (ksi)
AS4	278	19.7	1.40	201	19.0	17.4
AS6	318	20.6	1.52	212	19.3	19.1
IM6	350	24.6	1.38	211	19.8	18.3
HMS4	167	31.2	0.53	—	—	10.7

Fiber Volume = 62%

lected graphite fibers. As can be seen, the high tensile strengths and moduli, and the low density of the fibers are translated into composites with the high specific strengths and moduli that make them so attractive for use in demanding applications such as primary aircraft structures.

In addition to these specific mechanical properties, graphite composites possess outstanding fatigue and creep resistance. They are also noted for their thermal expansion characteristics because they can be constructed so as to yield nearly zero coefficient of thermal expansion structures.

Applications. Graphite fiber reinforced composites have found applications in a variety of areas. Their major usage is in the aircraft/aerospace industry where their high performance characteristics are essential to a variety of demanding applications such as aircraft wings, empennage and fuselage structures, spacecraft, missiles, etc. Graphite composites have also found use in such areas as sporting goods (golf clubs, tennis rackets, fishing rods), agriculture, materials handling equipment, medical devices, weaving machines, and other industrial applications.

Aramid Fibers. *Background.* Aramid fibers are organic fibers that possess high strength and high stiffness as a result of highly aligned, rodlike polymeric molecules. The high strength and stiffness are combined with a relatively low density to give very high specific tensile properties.

The most important aramid fibers are du Pont's Kevlars®,* which were introduced in the early 1970s and soon found many applications where the high specific properties could be utilized to advantage. They were used quite often to replace glass fibers in many applications due to their higher specific moduli and handleability characteristics. They were not satisfactory as replacements

---

*Registered trademark of E.I. du Pont de Nemours & Company.

for graphite fibers in demanding applications such as primary aircraft structures because of their low compressive strength. However, due to their high toughness, they have been combined with graphite to form hybrid composites having improved damage tolerance.

Aramid (aromatic polyamide) fibers are produced by an extrusion/spinning process. The polymer, dissolved in strong acids such as concentrated sulfuric, is extruded from spinnerets into cold water, washed thoroughly, and dried on spools. The fiber properties can be varied by altering processing parameters such as spinning conditions and heat treatments.

The Kevlar fibers are believed [15] to be an aromatic polyamide, p-phenylene terephthalamide, with the chemical formula

$$\left[\begin{array}{c} \underset{\|}{O} \\ -C \end{array} - \bigcirc - \underset{\|}{\overset{O}{C}} - \underset{|}{N} - \bigcirc - \underset{|}{N} - \right]_n$$

The highly aromatic structure formed through all para chemical structures produces the rodlike, rigid molecules having high tensile properties. The processes used in producing the fibers result in the alignment of the rodlike molecules in the fiber's axial direction.

While the polymer chains are held together in the axial direction by strong convalent bonds, the chains are held together in the transverse direction by hydrogen bonding. Consequently, the fiber, like graphite, is anisotropic. It exhibits high axial strength and comparatively low transverse strength.

One characteristic which significantly affects the fiber's performance is its morphology. It has been suggested [16] that the fiber possesses a fibrillar microstructure. Studies [17] have shown that a tensile failure of the fiber is characterized by longitudinal fragmentation and splintering into subfilaments. Under compression loading, localized buckling of the fiber has been observed.

Fiber Properties. Typical properties of du Pont's Kevlar 49 fibers, the most important Kevlar fiber for advanced composites, are shown in Table 1-23. As can be seen, aramid fibers possess high tensile strengths and moduli combined with a low density, giving them very high specific tensile properties. They also possess good retention of these properties to elevated temperatures, good chemical resistance, and good fatigue resistance.

Reinforcement Forms. Kevlar 49 is available in a variety of product forms, including yarns, rovings, and fabrics. The filaments themselves are also available in a variety of deniers, and the yarns and rovings are available in different

Table 1-23. Typical Properties of Kevlar 49 Fibers [18]

Property	Value
Density, g/cc	1.44
Tensile strength, ksi	525.0
Tensile modulus, Msi	18.0
Breaking strain, %	2.5
Axial coefficient of thermal expansion 10^{-6} in./in. °F	−1.1
Decomposition temperature, °F	932
Long-term use temperature, °F	320
Fiber diameter, μm	11.9
Axial thermal conductivity, Btu in./hr ft^2 °F	0.285
Ambient moisture absorption at 72°F/55% RH, %	3.5

cross-sectional sizes depending on the number of filaments per strand. The fabrics also come in a variety of forms depending on yarn size, weave, etc.

Composites can be fabricated from a variety of starting materials such as preimpregnated tapes, fabrics, and rovings. Rovings can also be impregnated with resin *in situ* with the filament winding process. In general, most of the starting materials discussed for glass are available for the aramid fibers. As previously noted, the fabrication process would be dictated by the material form and specific resin system.

Composite Properties. Typical properties attainable with Kevlar 49 composites are shown in Table 1-24 for unidirectional, continuous fiber reinforced epoxy resin matrix composites. As can be seen in the table, the high tensile strengths and moduli and low density are translated into composites with high specific tensile strengths (4×10^6 inches) and high specific tensile moduli (2.2×10^8 inches).

In addition to these specific mechanical properties, Kevlar 49/epoxy composites also demonstrate excellent fatigue resistance, being comparable to boron/epoxy composites and superior to glass/epoxy composites [18].

Applications. Due to its high specific tensile properties and ease of handling, Kevlar 49 is well-qualified for filament winding of pressure vessels. A considerable amount of work has been performed in this area, and substantial history and data have been established. For example, work at Lawrence Livermore Laboratory [5] showed that hoop-fiber stress values as high as 450 ksi could be obtained from Kevlar 49/epoxy composite pressure vessels.

In addition to pressure vessels, Kevlar composites have found applications in a wide variety of areas such as aerospace, leisure, protective clothing, and automotive.

Table 1-24. Properties of Unidirectional Kevlar 49/Epoxy Composites [18]

Property	Value
0° Tensile	
Strength, ksi	200.0
Modulus, Msi	11.0
Strain-to-failure, %	1.8
90° Tensile	
Strength, ksi	4.3
Modulus, Msi	0.8
Strain-to-failure, %	0.6
0° Compression	
Strength, ksi	40.0
Modulus, Msi	11.0
Density, g/cc	1.38
Fiber content, vol %	60
0° Flexure	
Strength, ksi	90.0
Modulus, Msi	11.0
Interlaminar shear strength, ksi	6-12
Dielectric constant	3.29
Axial Coefficient of thermal expansion, 10^{-6} in./in. °F	-2.2
Axial thermal conductivity, Btu ft/hr ft^2 °F	1.0

Another important use of Kevlar has been in hybrid composites. It can be combined with other advanced composite reinforcements to achieve an optimum balance of properties. For example, it could be incorporated into graphite/epoxy composites to yield a tougher, more damage resistant structure. It could also be combined with glass to give a lighter weight, stiffer structure. A wide variety of constructions are possible.

Miscellaneous Fibers. In addition to those fibers previously discussed, there are many other important fibers available which can produce composites that will meet a variety of specialized applications. Some of these fibers will be briefly discussed.

Fiber FP. Fiber FP is an alumina fiber manufactured by du Pont whose composition is > 99% α-alumina. It is prepared as a continuous yarn and is suitable for reinforcement of organic resin, metal, and ceramic matrices. It is available in yarns, tape, and fabrics. Typical properties of Fiber FP are shown in Table 1-25. Notable among its characteristics are excellent chemical resistance, very high modulus, good electrical insulation, and very high temperature stability.

Table 1-25. Typical Properties of du Pont's Fiber FP [19]

Property	Value
Tensile strength, ksi	200
Tensile modulus, Msi	35
Density, g/cc	3.9
Melting point, °F	3713
Filament diameter, μm	20

Table 1-26. Typical Properties of Nextel 312 Fibers [20]

Property	Value
Tensile strength, ksi	250
Tensile Modulus, Msi	22
Density, g/cc	>2.7
Continuous use temperature, °F	2200
Short-term use temperature, °F	2600
Filament diameter, μm	
900 denier	10–12
600 denier	8–9
Metal oxide components, wt %	
Al_2O_3	62
B_2O_3	14
SiO_2	24

Nextel®.* Nextel ceramic fibers are continuous, polycrystalline metal oxide fibers from 3M. The fibers are available in a variety of product forms, including rovings, yarns, and fabrics, and can be used to reinforce a variety of matrix materials. Typical properties of one type of Nextel fiber, Nextel 312, are shown in Table 1-26. These fibers are characterized by high temperature stability, very good electrical insulation, very good thermal insulation, and excellent abrasion resistance.

Nicalon®.† Nicalon fiber is a silicon carbide fiber manufactured by Nippon Carbon Company, Ltd., of Japan and distributed in the United States by Dow

---

*Registered trademark of 3M Company.
†Registered trademark of Nippon Carbon Company.

Table 1-27. Typical Properties of Nicalon Fibers [21]

Property	Value
Tensile strength, ksi	360–470
Tensile modulus, Msi	26–29
Strain-to-failure, %	1.5
Axial coefficient of thermal expansion, $10^{-6}/°C$	3.1
Diameter, μm	10–15

Corning Corporation. It is composed of ultrafine β-SiC crystals with excess carbon. It is available in a variety of product forms, including continuous fiber, fabrics, braids, mats, and ropes, and is suitable for reinforcing a variety of matrix materials, including organic, metals, and ceramics. Typical properties of Nicalon fibers are shown in Table 1-27.

Nicalon fiber characteristics include high temperature stability (50% retention of tensile strength at 1200°C), very good chemical resistance, good thermal insulation, and non-electrical conductive.

References

1. Lee, H., and K. Neville, *Handbook of Epoxy Resins*, McGraw-Hill, New York, 1967.
2. Russell, S. E., "AS6 and Other New Hercules Carbon Fibers," Orange County SAMPE Seminar, February 1983.
3. Edelman, R., and P. E. McMahon, "A New DAP-Polyester Resin for Carbon Fibers," *Composite Technology Review*, **1,** No. 2, 7, 1979.
4. *Modern Plastics Encyclopedia*, **60,** 10A (October 1983).
5. Lubin, G., Ed., *Handbook of Composites*, Van Nostrand Reinhold, New York, 1982.
6. Serafini, T. T., et al., U.S. Patent 3, 745, 149, July 1973.
7. Serafini, T. T., "Status Review of PMR Polyimides," *ACS Organic Coatings and Plastics Chemistry*, **40,** 469 (April 1979).
8. Delmonte, J., *Technology of Carbon and Graphite Fiber Composites*, Van Nostrand Reinhold, New York, 1981.
9. Street, S. W., "V-378A, An Addition Polyimide for Advanced Composites," First International Conference on Polyimides, Society of Plastics Industry, Ellenville, N.Y., November 1982.
10. Fulmer, R. W., "S-2 Glass Bridges a Gap in the Reinforcements Spectrum," National SAMPE Symposium, 1980.
11. Technical Brochure, Avco Corporation.

12. *Advanced Composites Design Guide*, Third Edition, Second Revision, Air Force Flight Dynamics Laboratory, Wright-Patterson Air Force Base, Ohio, 1976.
13. Union Carbide Corporation, *Carbon Fiber Products Catalogue*, September 1983.
14. Hercules Incorporated, Magnamite® Graphite Fiber Brochure, 1981.
15. Penn, L., H. A. Newey, and T. T. Chiao, "Chemical Characterization of a High Performance Organic Fiber," *J. Mat. Sci*, **11,** 190 (1976).
16. Abbot, N. J., et al., "Some Mechanical Properties of Kevlar and Other Heat Resistant, Nonflammable Fibers, Yarns, and Fabrics," AFML-TR-74-65, Pt III, 1975.
17. Chiao, T. T., et al., "Filament Wound Kevlar 49/Epoxy Pressure Vessels," Lawrence Livermore Laboratory Report UCRL-51466 (1973).
18. *Kevlar® 49 Data Manual*, du Pont, 1977.
19. Technical Brochure, "Fiber FP Fiber, Tape & Fabrics," du Pont, 1978.
20. Technical Brochure, "Nextel® Ceramic Fiber," 3M Company.
21. Technical Brochure, "Nicalon® Silicon Carbide Fibers," Dow Corning, 1983.

2
FABRICATION TECHNIQUES
Review and Perspective for Application Engineers

W. Brandt Goldsworthy
President
Goldsworthy Engineering, Inc.
Torrance, California

Composite fabrication techniques originally consisted exclusively of manual labor, where laminators typically laid reinforcing mat into a plaster mold and applied resin. Although a few production-oriented companies utilized metal molds that were heated and used in standard presses, most fabricators achieved atmospheric pressure by vacuum bagging, and cured the part at room temperature or in an oven.

An astonishing number and variety of first applications for FRP materials came into being by such hand lay-up methods, many of which have survived to this day, including boat hulls, wings and fuselage for military aircraft, molded chair bottoms, aircraft radomes, special purpose pressure pipe, corrugated translucent panels, sports car body shells, chemical processing tanks, and marina floating wharfs. Some of these long-term reinforced plastic/composite (RP/C) applications are still commonly produced by traditional hand methods; others, such as tanks, pipe, and panels, are regularly manufactured on production equipment.

In the early 1950s, as the diversity and volume of applications for these materials became apparent, the industry began seeking ways to automate and develop true composite production systems. One of the first inventions resulting from this effort was a glass fiber preform machine that sprayed chopped glass strands onto a three-dimensional air screen which approximated the shape to be molded. The preformed glass strand mat was then placed within a matched-metal die in a regular molding press, and liquid resin was added to form the product. Such preform/die-molding processes have been used to mass produce automobile bodies, truck cabs, free-form furniture, boxes and trays, reusable concrete forms, and small boat shells.

Another significant advance was accomplished when hand-made tubular glass fishing rods were replaced by rods produced in volume on automatic taper rod rolling machines and affiliated handling equipment.

These and other preliminary attempts to automate hand processes led ultimately to the development of four generic automated composite manufacturing systems: i.e., filament winding, pultrusion, pulforming, and tape placement, with other specialty equipment custom designed for specific applications.

In this chapter, we will deal only with the generic systems; reference to bibliographical material will acquaint the reader with fabrication techniques developed for specialty applications.

FILAMENT WINDING

The formative years of filament winding as a composite fabrication technique were characterized by high precision and rapid growth, but low volume production. This circumstance evolved primarily as a result of the fact that, although filament winding was born into the pipe and tank industries in the early 1950s, it was quickly adopted by the then burgeoning aerospace industry for products with high strength:weight ratios. Aerospace/military requirements placed ultra-stringent precision, repeatability, and quality control demands on material and equipment manufacturers, who set and maintained high standards in order to meet these requirements. At the same time, the political climate of the times gave military spending a high priority, which allowed filament winding development to proceed at a far more rapid rate than it might have otherwise, but at the same time limited its transference to commercial, high volume usage.

Some common types of filament winding and typical equipment are described and illustrated in the following sections.

Lathe Type (Figure 2-1)

Early filament winding machine models were generally converted engine lathes, with the product mold, or mandrel, turning, while the material supply (resin-wet roving) traversed the length on a fixed path and plane. The product thus wound is one of simple geodesic paths, instead of the complex geometry available with today's computer controls.

A typical lathe winder is illustrated in Figure 2-2. The equipment shown, built in the mid-sixties, was dedicated to winding the motor case directly on

40 ADVANCED THERMOSET COMPOSITES

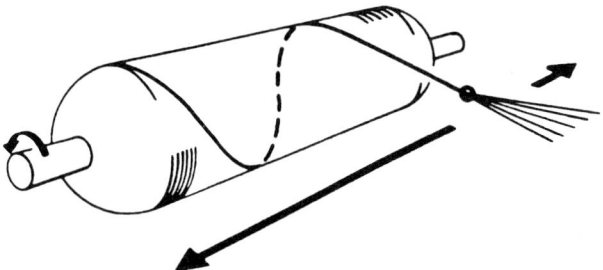

Figure 2-1. Lathe-type winder.

cast live propellant as a mandrel and utilized numerical tape controls with remote operation capability—an advanced concept at that time.

Figure 2-3 is a recent, computer controlled, multi-axis, general purpose, lathe-type filament winder. It is designed for modular expansions of the traverse length and centerline height. A computer numerical control (CNC) system allows the user to tailor a control package to his immediate needs while also facilitating expansion as future needs require. Three computer controlled axes of motion provided on the traverse assembly—i.e., traverse (X-axis) motion, crossfeed (Z-

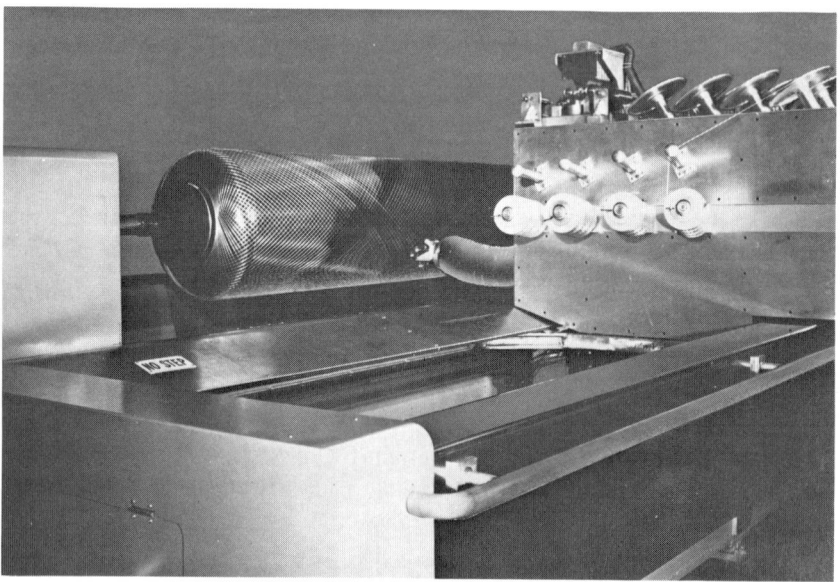

Figure 2-2. Typical lathe-type filament winder—NC controls with remote for winding on cast live propellant.

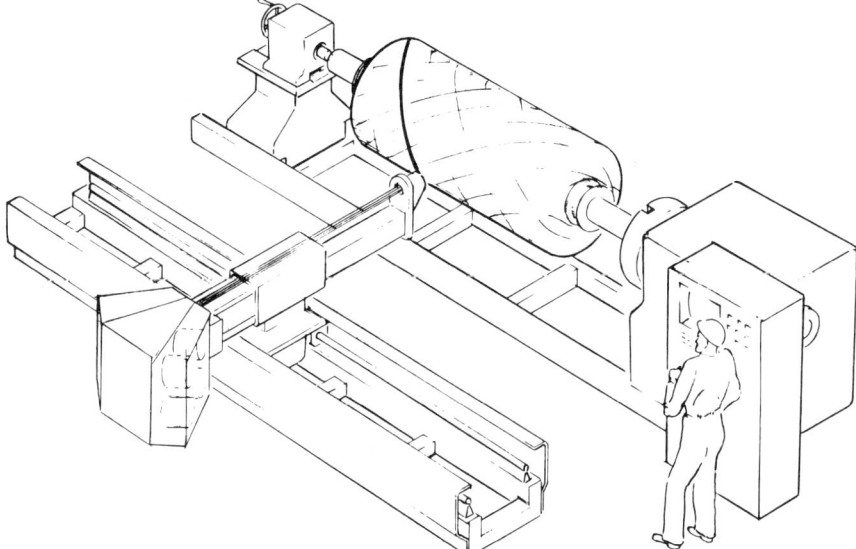

Figure 2-3. Computer controlled, multi-axis, general purpose lathe-type filament winder.

axis) motion, and feedeye rotation—enable winding of circumferential and helical patterns, as well as conical shapes. The key feature of this delivery mechanism is that the feedeye rotates around the fiber band centerline, rather than describing an arc at the point of end-of-mandrel dwell, which is the more common method. The catenary problem created by the standard feedeye as it arcs, and the resultant tension variation, are virtually eliminated in this design, with resultant laydown accuracy. The process may be operated either with a wet resin system or with preimpregnated reinforcements.

The helical design commonly associated with filament wound pipe and tank is a natural outcome of the lathe-type process, and led to the erroneous impression that all pipe and tanks must be helically wound, with the result that when newer more efficient processes incorporating longitudinal/circumferential winds were later introduced for these products, they were viewed with suspicion. However, recognizing their superior efficiency, the aerospace industry soon began utilizing these new techniques for high performance motor cases and tankage.

Whirling Arm (Figure 2-4)

While lathe-type filament winders continue to be used to produce a variety of cylindrical structures, including those with integral closed ends, a shortcoming

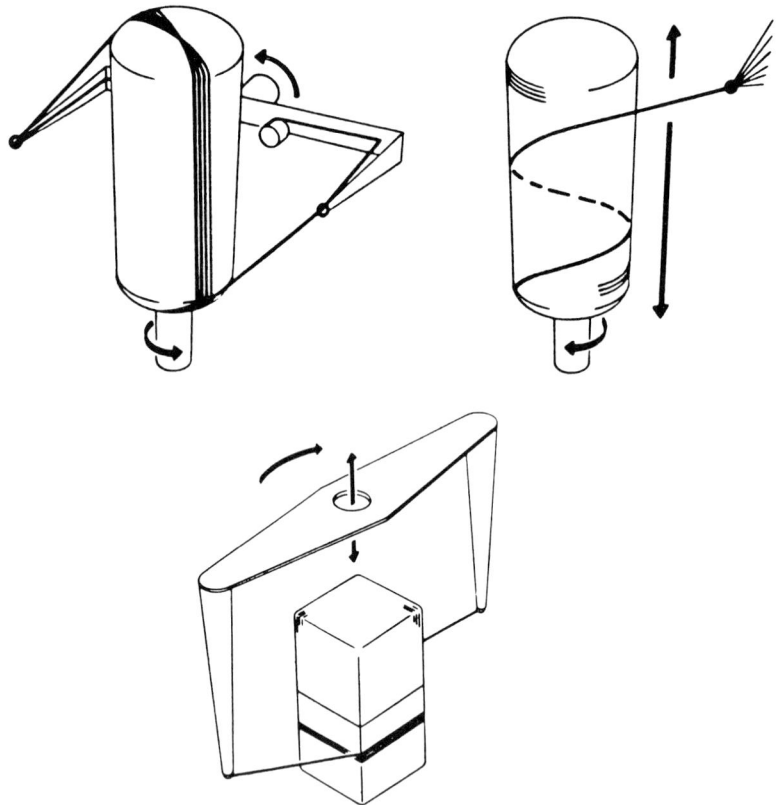

Figure 2-4. Whirling arm–type winder.

of the process is that it is extremely slow to lay down polar longitudinals, as the mandrel must be indexed 180° at the end of each path. This factor led to development of the so-called whirling arm–type filament winding machine, which was first developed to wind high efficiency pressure vessels for aerospace use where true longitudinal filament placement was essential. The principle was soon broadened to encompass winding nonrotational surfaces for producing high strength instrument cases. A box winding machine built in the early sixties wound six faces of the tool, utilizing hydraulically operated tables to flip the tool into position (see Figure 2-5).

Racetrack Type (Figure 2-6)

Because there was some difficulty in winding in all three modes (polar, circumferential, and helical) with the whirling arm–type equipment, the racetrack ma-

FABRICATION TECHNIQUES 43

Figure 2-5. Box winding machine—Whirling arm filament winding machine for manufacture of rectangular shapes.

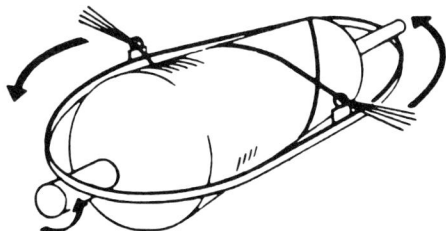

Figure 2-6. Racetrack-type winder.

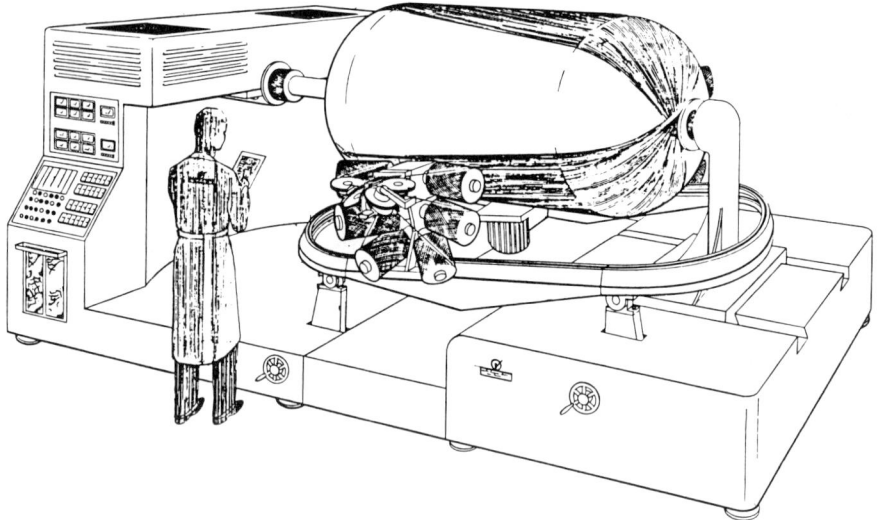

Figure 2-7. Typical racetrack-type filament winder.

chine came into being. In conjunction with its unique capability of winding in all these conventional modes, the racetrack-type winder incorporates the additional advantage of having two pay-on carriages antipodally disposed in directly opposed locations around the track, thus applying twice the amount of reinforcing strands per cycle. A typical racetrack machine is illustrated in Figure 2-7. The equipment shown is fabricating a rocket motor case, which is a common closed-end pressure vessel application of this process.

Tumble (Polar Orbital) Winder (Figure 2-8)

Another basic winding principle offering essentially the same functional advantage as the racetrack winder is the tumble, or end-over-end, winder, wherein longitudinal winding is achieved by rotating the mandrel through its longitudinal axis. Helical and circumferential winding are normally done by locking the spindle in a fixed position to parallel the motion of an ancillary reciprocating carriage. Simplicity of construction and motion make this principle useful in small machines where precision and versatility are paramount. As the mandrel must be cantilevered, this machine is somewhat size-limited, and mode selection mechanics are complex for most production requirements. A laboratory and research winder built on this principle is illustrated in Figure 2-9.

A small, low cost production tumble winder with cantilevered mandrel is illustrated in Figure 2-10. Larger production machines use a gantry traverse

FABRICATION TECHNIQUES 45

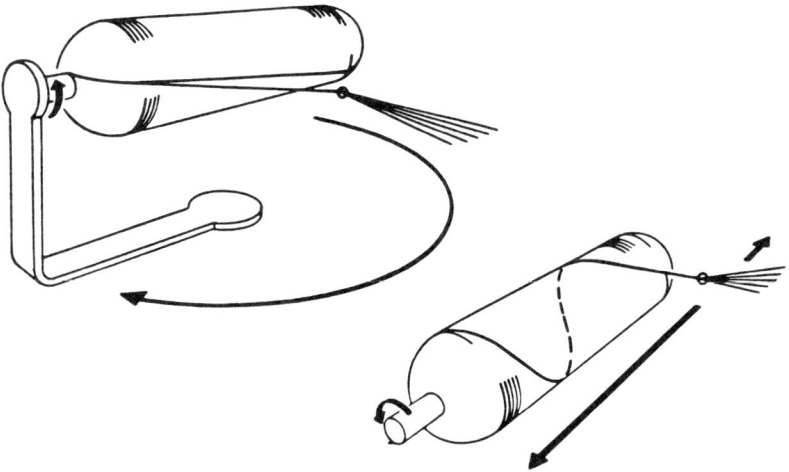

Figure 2-8. Tumble-type winder.

Figure 2-9. Tumble-type laboratory/research filament winder.

46 ADVANCED THERMOSET COMPOSITES

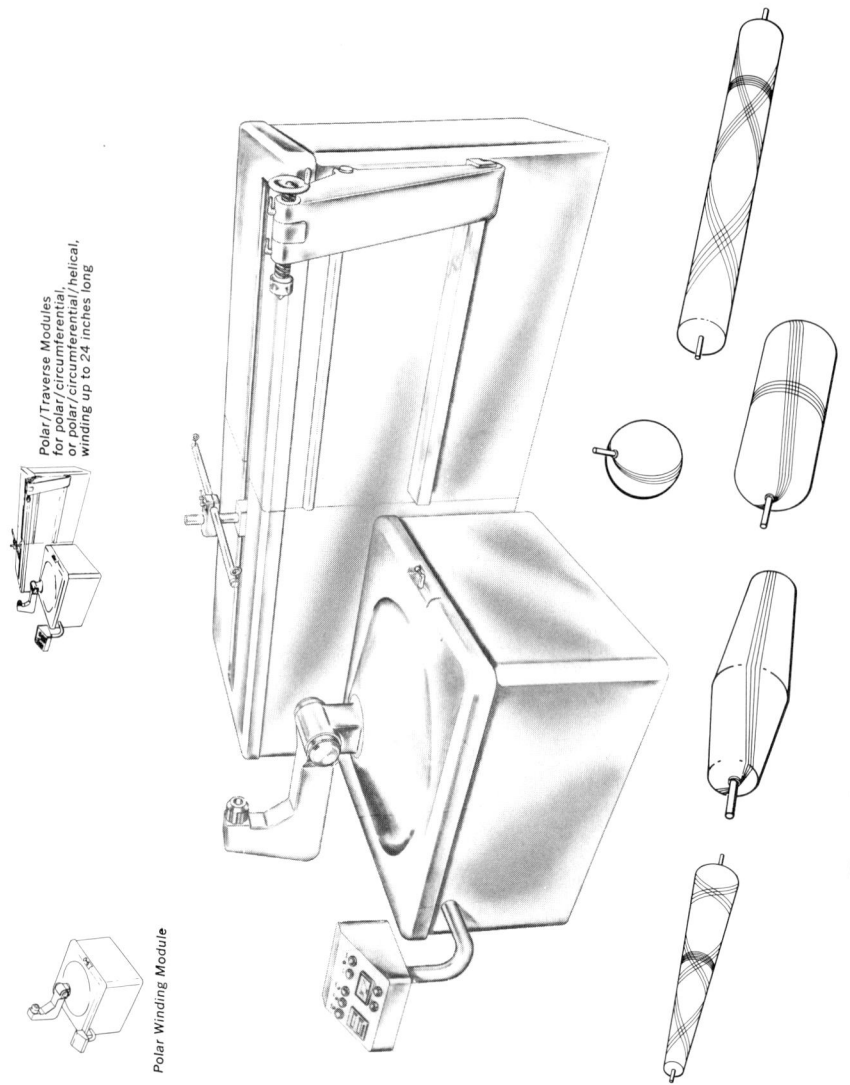

Figure 2-10. Typical small production tumble-type winder.

system, as in Figure 2-11, to assure precision in the helical and circumferential modes under heavy filament loads.

Ring Winder (Figure 2-12)

Massive structures such as helicopter and windmill blades can be produced on ring winding equipment. The size of such articles mandates a stationary mandrel. The tool is therefore fixed between headstock and openable tailstock, and further supported by steady rests stationed along the length of the bed. A winding wheel forms a ring around the mandrel and is loaded with the required roving or tow. It revolves as it reciprocates along the tool length, laying a preprogrammed pattern of longitudinal, helical, and/or circumferential strands. Steady rests are hydraulically lowered and raised to accommodate passage of the ring. (see Figure 2-13)).

Spherical Winder (Figure 2-14)

One of the most difficult types of winding is spherical winding. The process is made even more challenging when the product is one of the contact-type balls—e.g., basketballs, volleyballs, and soccer balls—all of which are filament wound. The intrinsic paradox of this process is that while a totally random wind must be effected, it must also be a *controlled* totally random wind, and because it is necessary to wind everywhere on the ball, there is no way of holding on to it to drive it. One solution to this dilemma is to suspend the rubber bladder on an air column; product rotational speed, and thereby direction of wind, can be changed by air jetting the column, thus achieving the requisite controlled random pattern. Prototype equipment of this type is shown in Figure 2-15.

PULTRUSION

Another generic automated process developed for the FRP industry is pultrusion, which also made its appearance in the 1950s and has evolved into a number of variations, most of which today utilize programmable controls.

Pultrusion is the one-step conversion of raw composite materials to finished structural product, continuously. It is similar, but not identical, to the conventional aluminum extrusion process, with the variation that a number of raw composite materials enter the processing equipment simultaneously and are pulled (hence, the term "*pul*trusion") rather than pushed through the system.

The process consists, essentially, of those elements depicted in Figure 2-16. Resin-wet reinforcements are drawn into the sytem through squeezeout bushings, which remove excess resin, and are optionally preheated by dielectric radio frequencies (5–100 MHz) or microwave frequencies (915–2450 MHz) or,

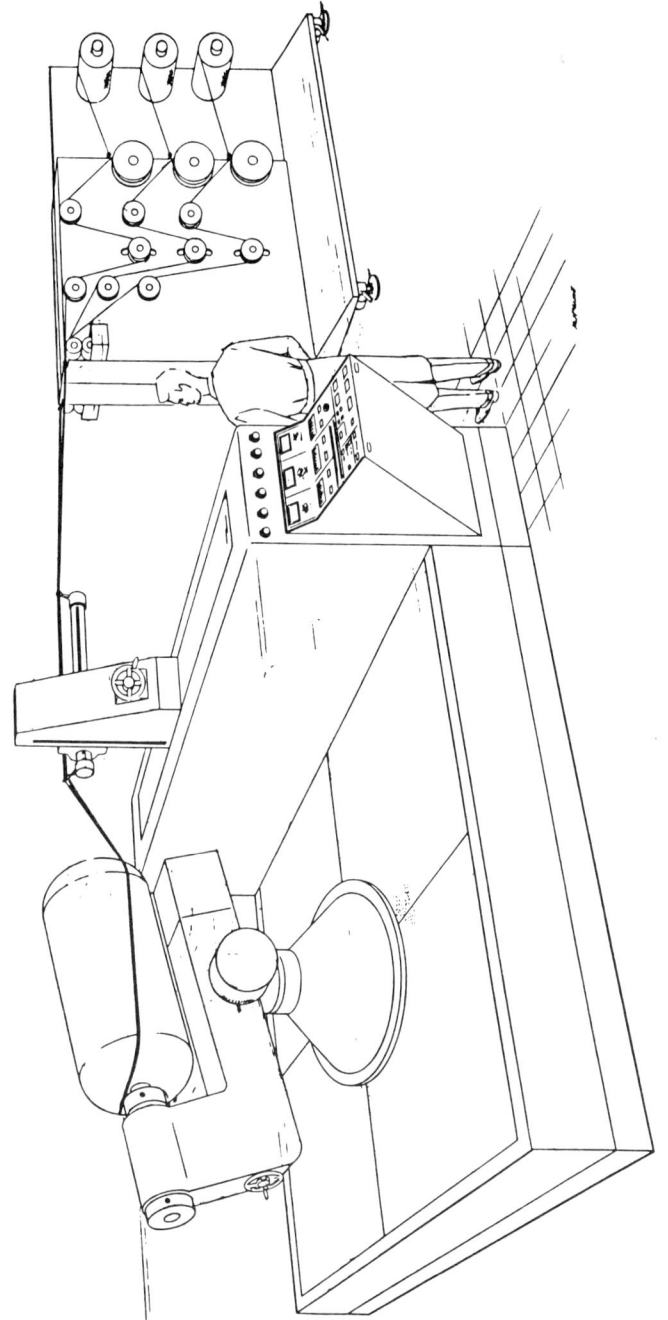

Figure 2-11. Production tumble-type winder with gantry traverse for larger products.

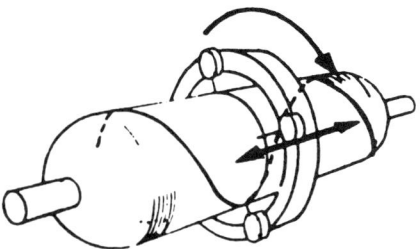

Figure 2-12. Ring-type winder.

in cases where conductive reinforcements such as graphite are included, by induction preheating. The roving package then enters a heated forming/curing die, and the cured stock is pulled out of the die by suitable pulling devices.

In heavy sections, throughput rates were dramatically increased when high frequency preheat was introduced to pultrusion equipment (Figures 2-17, 2-18, 2-19). Experimentation ultimately led to the augmented cure invention; this optimum synergistic combination of cure methods first achieves instantaneous preheat throughout the fiber/resin mass, and then pulls the mass immediately into a heated steel die (see references 7, 8, and 9 for patents). Since the material will already have been exposed to sufficient energy to achieve cure during the preheat phase, heating the steel die is really more a means of preventing cure heat loss than a means of introducing additional heat.

The first production application for the pultrusion process and equipment was power pole lanyards for the electrical industry. Pultruded products soon covered a wide range in the tremendously broad market for electrical bar stock that is characterized by unidirectional reinforcement and constant cross section. However, since pultrusion was initially conceived as a purely unidirectionally reinforced process, the industry was excluded from markets that required products with off-axis fiber orientation for structural purposes.

Off-axis Fiber Orientation

In the early seventies, the process was revolutionized when in-line infeeding of multidirectional reinforcement was accomplished, thus enabling pultrusions with off-axis structural properties. The application that brought about this innovation was square tubing, to be used for purposes requiring greater structural properties than unidirectional fiber could provide. The resulting equipment is shown in Figure 2-20; a tubular tooling line, convolutely folded mat, and near 90° circumferential windings are combined with pure longitudinal roving in a continuous pultrusion application to produce these high strength square tubes.

50 ADVANCED THERMOSET COMPOSITES

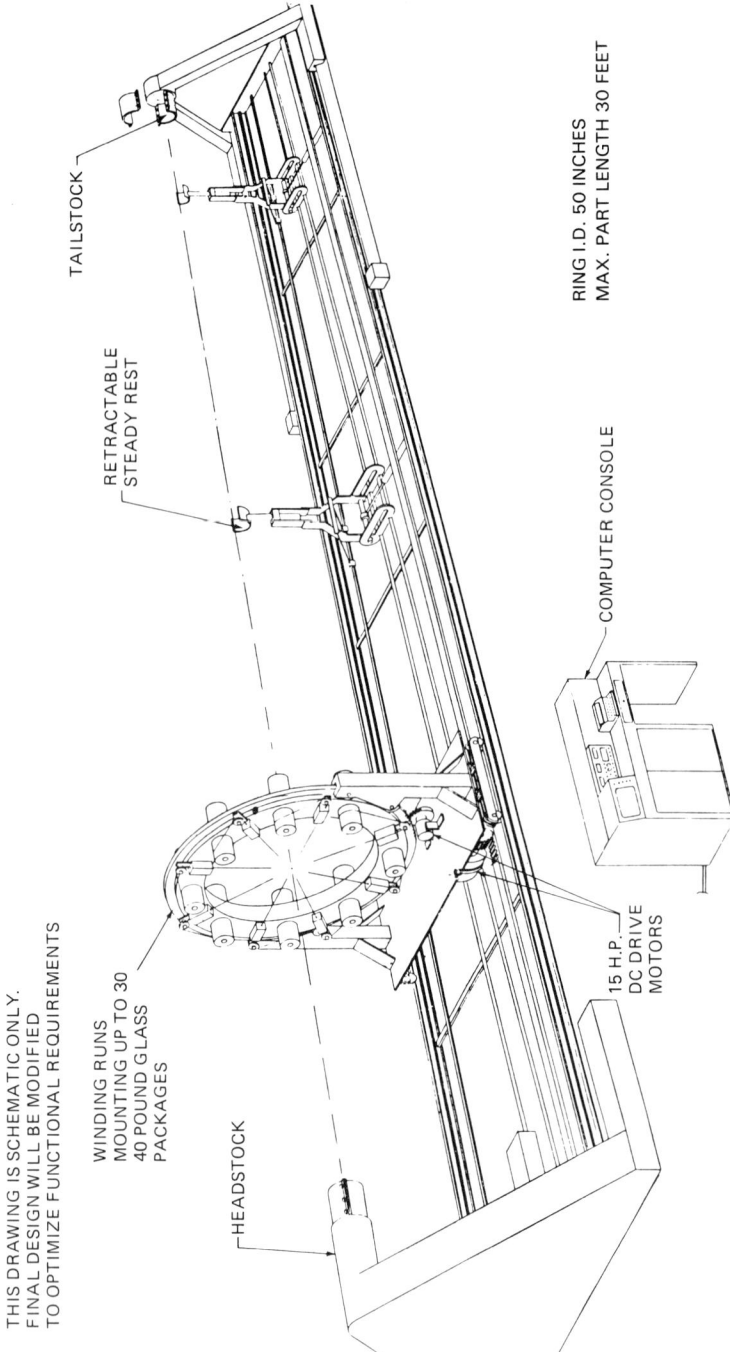

Figure 2-13. Typical ring-type filament winder.

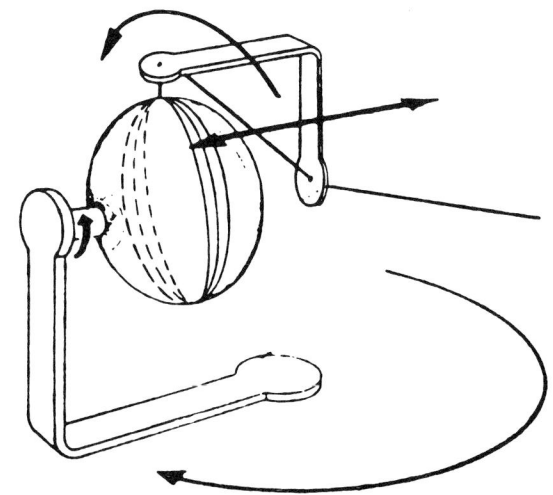

Figure 2-14. Spherical-type winder.

Figure 2-15. Spherical filament winder with rubber bladder suspended on an air column.

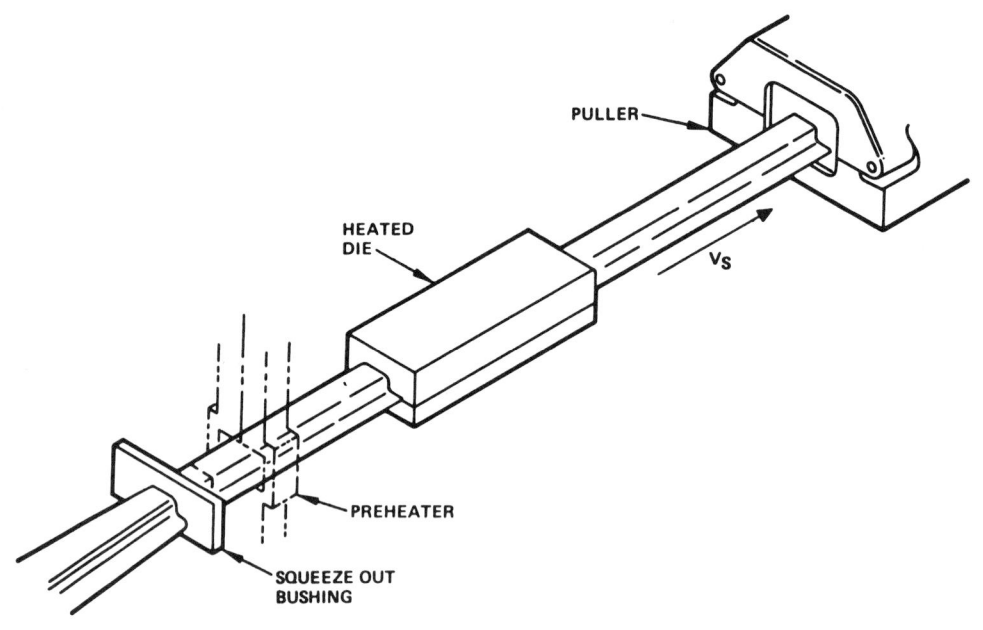

Figure 2-16. Pultrusion process.

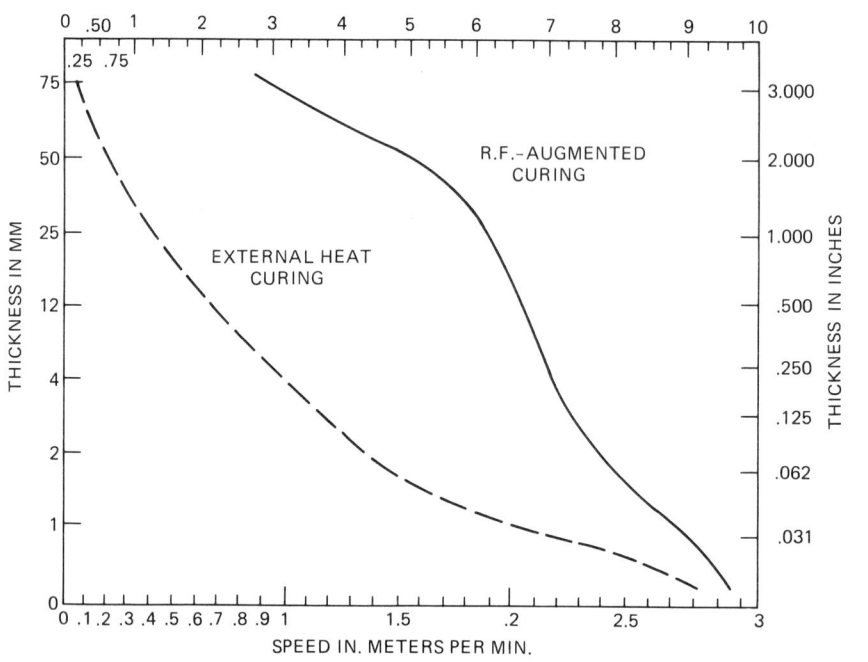

Figure 2-17. RF augmented cure rate—graph 1.

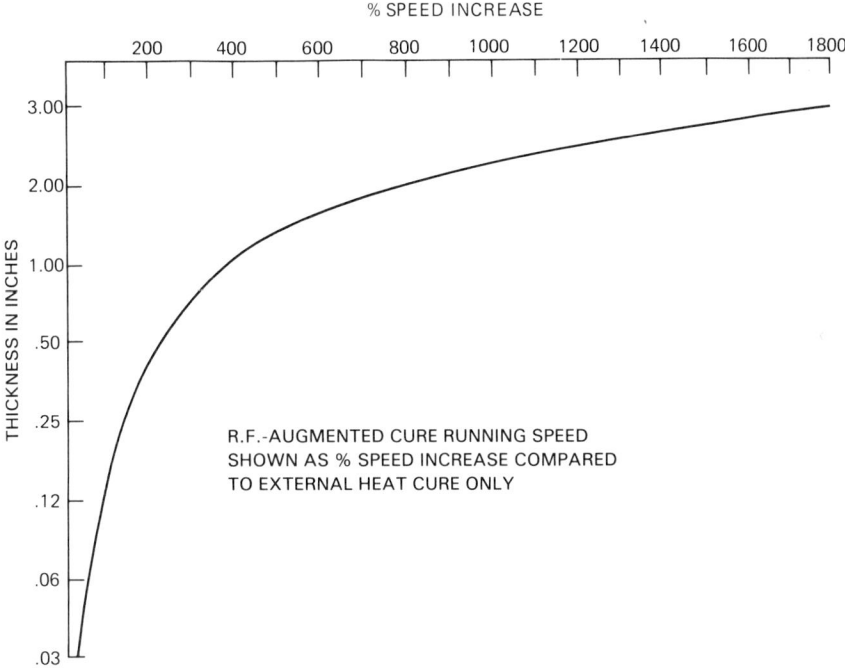

Figure 2-18. RF augmented cure rate—graph 2.

This successful venture provided the impetus that led to the tube manufacturing equipment illustrated in Figure 2-21, in which it was proven that virtually any fiber orientation and laminate schedule could be pultruded to produce stock with structural properties that equal equivalent traditional metal stock. The tubing construction starts with a combination of an inside ply of polyester veil mat brought into the system through a top-side spool, a ply of random glass fiber, and a ply of longitudinal fibers. Because of the thick wall, this initial material package is then taken through an impregnating bushing. After the longitudinals and first two plies of veil and mat are impregnated, another ply of random glass mat is added, and a ply of pure circumferentials are wrapped around the package by a winding wheel. Another ply of random mat is added, following which a +45° ply and a −45° ply are added by large winding wheels. Another impregnating bushing wets out this last group of plies. The package than enters the radio frequency (RF) system, where all the energy is introduced. (After this tooling had been designed and built, the customer requested another ply of random fiber mat and a ply of C glass—chemical glass—on the outside. Therefore, while it is not recommended, these plies were added after the fact down-

54 ADVANCED THERMOSET COMPOSITES

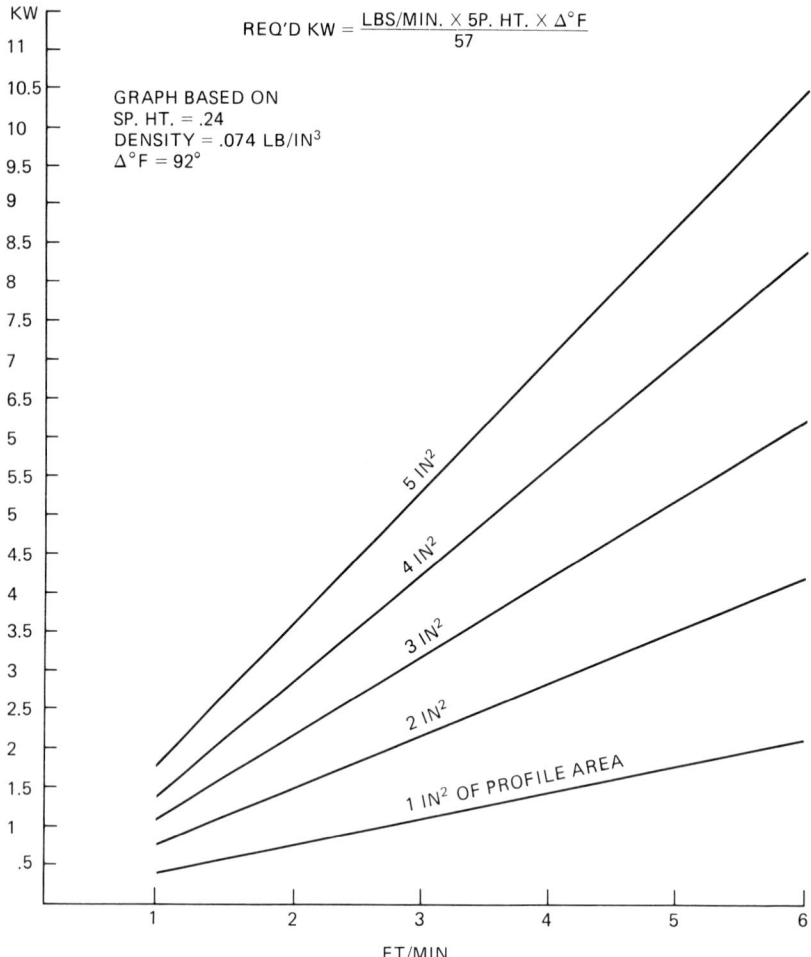

Figure 2-19. RF power required for augmented cure.

stream of the RF cabinet.) The product than proceeds through the pullers in the usual manner.

Materials

Essentially every accepted form (roving, tow, mat, cloth, braiding), from all types of filaments (glass, graphite, Kevlar®,* boron, steel, nylon, etc.) may be used for pultrusion. The most common raw materials are fiberglass roving with a wet polyester resin system, and most of the standard pultrusion equipment

---

*Registered trademark of E. I. du Pont de Nemours & Company.

FABRICATION TECHNIQUES 55

Figure 2-20. Pultrusion equipment for square tubing with off-axis reinforcement.

Figure 2-21. Pultrusion equipment for round tubing with multiple ply laminate schedule and multi-directional reinforcement.

applications listed below are of this makeup. However, any of the forms and types of material may be adapted to off-axis tooling and/or utilized in typical standard equipment.

Typical Standard Production Equipment

Figure 2-22 illustrates a current generation, standard 8 × 24 inch (20 cm × 60 cm) capacity pultrusion production machine. Machine motions are controlled by a programmable controller and are hydraulically driven. The flying cutoff saw, an integral part of the system, is synchronized with the reciprocating gripper/puller speed and motion.

Standard Equipment Applications

The list of current product applications from such typical, standard capacity pultruders is broad and high volume, but represents only a portion of the total pultrusion market when one considers the variant processes arising out of the basic pultrusion technique, some of which will be discussed later in this chapter. Standard pultruded product includes the following:

1. Electrical industry—motor and transformer pultruded parts (e.g., slot wedges, winding pins, transformer spacers or insulators, etc.), antenna guy lines, strain and suspension insulator rods, switch lanyards, and so forth.
2. Sporting goods and recreational industry—solid fishing rods, recreational tent ribs, sail battens, recreational vehicle warning antennae, ski facings and ski poles, drive-belt cleats on snowmobiles.
3. Miscellaneous—door rocker panels for cars and trucks; side rails for lightweight, nonconductive, weatherproof ladders; marine and truck windshield channels; corrosion-resistant safety hand railings; tool and hammer handles; refrigerated wall spacers and structural members; underground conduit.

Enlarged Capacity Pultruders

Because the pultrusion process is essentially a no-pressure process, there is no technical limit to the size of profile that can be produced. Consequently, the largest marketing area of all may be in pultruded profiles that are substantially larger than any profiles that can be produced in metal. An example of this trend to larger equipment is the 18 × 36 inch (45 × 90 cm) capacity pultrusion machine shown in Figure 2-23. The upstream tooling provides the flexibility of adding mat or other, off-axis reinforcement to the longitudinal fiber-reinforced

FABRICATION TECHNIQUES 57

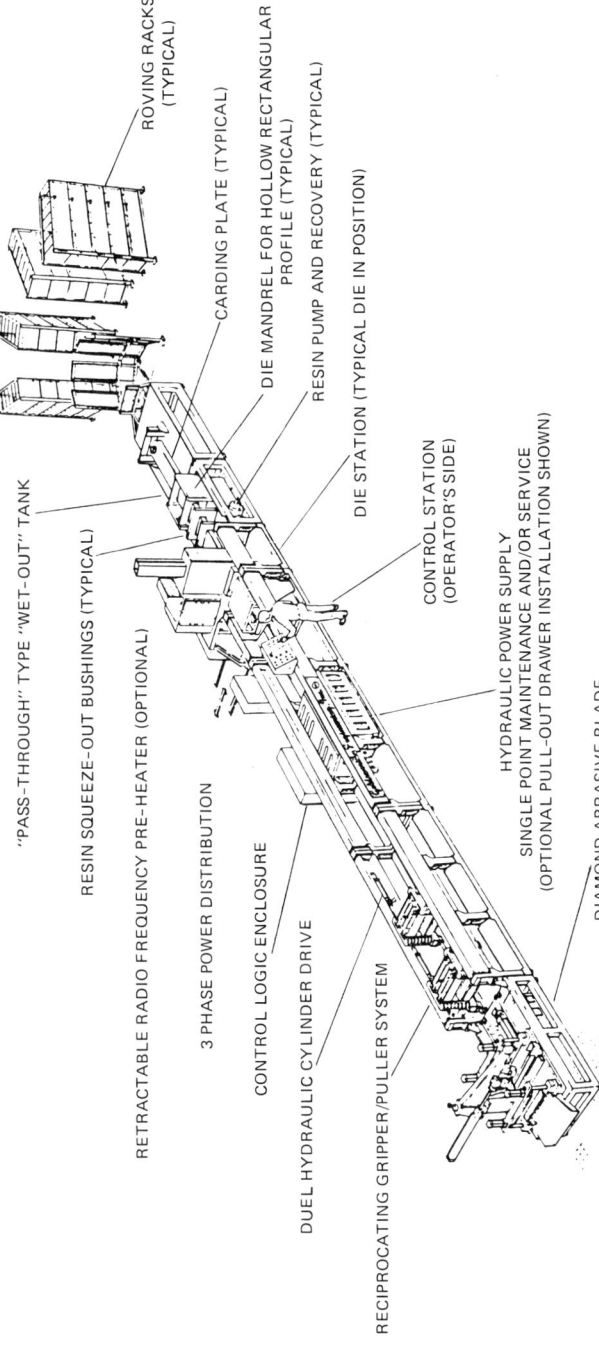

Figure 2-22. Typical standard capacity 20 × 60 cm pultrusion production machine.

58 ADVANCED THERMOSET COMPOSITES

Figure 2-23. Enlarged 45 × 90 cm capacity pultrusion machine.

plastic (FRP) stock. This large capacity unit is currently producing continuous stock consisting of multiple hollow sections with intercostal webs for the automotive industry.

New Directions

While space limitations preclude more than a cursory look at new trends in pultrusion, several are worthy of mention, and a few of these will be reviewed briefly.

1. The advent of exotic advanced composite materials such as carbon graphite reinforcements and fast cure epoxies has impacted pultrusion as well as other processes, and led to development of aircraft and aerospace applications. Pultruded structures appear to be cost effective as angles, tees, and hat stiffeners for skin panels, floor beams on commercial aircraft, spar sections for helicopter tail rotor blades, and hat stiffeners for missile inter-stages. As shown by Figure 2-24, graphite pultrusions are both weight and cost effective. While principles of operation are the same, such factors as enlarged radius redirects and close

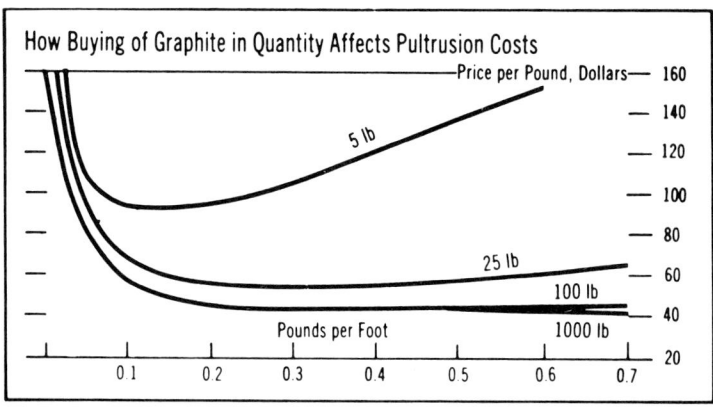

Figure 2-24. Graphite pultrusions—weight and cost data. (Reprinted from *Iron Age*, May 10, 1976)

temperature control must be considered when processing high modulus fibers and specialty resins.

2. Preformed pultrusions are another recent innovation which can produce pliant B-staged sections that can be stored in a freezer for later hand forming to fit a specific shape, with subsequent co-cure. Two recent programs have produced a curved and twisted helicopter door track utilizing E and S glass reinforcement, and a twisted turbine exit guide vane using graphite. Figure 2-25 illustrates the tooling setup for this process: various combinations of prepreg roving and cloth, along with dry roving or tow, are fed through two short heated dies that densify and debulk the laminate, with the tackiness of the prepreg bonding the preform together to make it handleable, without actually curing or even advancing the "B" stage. The resulting pultruded preform can be

60 ADVANCED THERMOSET COMPOSITES

Figure 2-25. Helicopter door track pultruded preform material exiting the second densification die.

put in cold storage or even shipped across country packed in dry ice for final molding at some later destination. When the part is to be molded, it is removed from cold storage, brought to room temperature, and placed in the mold along with the part it is to be attached to. The preform is generally then co-cured as an integral part of the assembly, usually in an autoclave. Or, as in the case of the guide vane, the preform is individually compression molded into the vane.

3. A considerable amount of work has also been done in the area of the thermoplastic matrix pultrusion involving most of the available high performance polymers. This seems to be a natural for thermoplastics in that it is a continuous process where the heat/chill cycles can be precisely maintained. The primary area of change between thermoset and thermoplastic pultrusion is in the die section. Where thermoset pultrusion is done with a single long die, with polymerization taking place entirely within the residence time in the die, it appears that thermoplastic pultrusion is best done with segmented dies providing a preheat melt-and-chill complex. Except for this die section change and puller speeds, existing pultrusion equipment seems well suited to thermoplastic pultrusion without extensive modification. An interesting post-processing capability provided by the thermoplastic matrix in the pultrusion market is the ability to postform within certain constraints and to join profiles by welding.

Basic Pultrusion Design Properties

Development of property tables for pultruded products can be misleading, as almost any required set of physicals can be tailored. However, the following table may be useful as a general guideline.

Basic Pultrusion Design Properties

Ultimate Strength—For preliminary design only

I. Fiber properties—E glass fiber at 72°
   A. Tensile strength   500,000 psi
   B. Tensile modulus 10.5×10^6 psi
   C. Specific gravity   2.54

II. Cured polyester resin—neat resin casting (iso-polyester at 72°F)
   A. Tensile strength   6000–10,000 psi
   B. Tensile modulus $3.0–6.4 \times 10^5$ psi
   C. Flexural strength   8500–18,300 psi
   D. Compressive strength   13,000–36,500 psi
   E. Specific gravity   1.10–1.46

III. Composite—unidirectional properties (at 72% by weight fiber)
   A. Tensile strength 0°
      (longitudinal)   160,000 psi
   B. Tensile strength 90°
      (transverse)   5000 psi
   C. Compressive strength
      0°(longitudinal)   85,000 psi
   D. Interlaminar shear strength   12,000 psi
   E. Modulus of elasticity 4.9×10^6 psi
   F. Density   0.075 lb/in.^{3}

IV. Composite—mat/polyester properties (at 30% by weight fiber)
   A. Tensile strength   14,100 psi
   B. Flexural strength   26,400 psi
   C. Compressive strength   27,600 psi
   D. Modulus of elasticity 0.99×10^6 psi

Rule of thumb: 3 oz of mat typical per 1/8-in. of profile wall thickness.

This laminate schedule achieves transverse strength and provides sufficient bulk for successful pultrusion.

Structural mat forms common to the industry are both 1 and 1-1/2 oz. A 1 oz mat with 70% by weight *resin* cures out at approximately 0.022 in. thickness. A 1-1/2 oz mat is approximately 0.030 in.

A 1/8 in. wall thickness pultruded profile will have a fiber content of approximately 50–52% by weight. The properties for such a profile are:

V. Combination of mat and unidirectional fibers for pultrusion of hollow profiles.

Longitudinal Direction

A. Tensile strength 25,000 psi
B. Compressive strength 30,000 psi
C. Flexural strength 37,500 psi
D. Tensile modulus 2.0×10^6 psi
E. Compressive modulus 2.3×10^6 psi
F. Flexural modulus 2.0×10^6 psi
G. Shear strength 6000 psi
H. Bearing stress 37,500 psi

Transverse Direction

A. Tensile strength 12,000 psi
B. Compressive strength 20,000 psi
C. Flexural strength 16,000 psi
D. Tensile modulus 1.5×10^6 psi
E. Compressive modulus 1.3×10^6 psi
F. Flexural modulus 1.3×10^6 psi

PULFORMING

Since there is no way to form or bend pultrusions after they are cured, if a curved pultrusion is needed, it must either be produced by B staging, as described earlier, or made initially in the curved configuration.

A curved version of the pultrusion process was first invented in the mid-seventies in response to NASA's requirement for graphite hat section stiffener rings for the shuttle vehicle. This initial development effort has matured into two separate but related fabricating techniques: (1) curved pulforming and (2) straight pulforming. Both of these processes have the capability of producing either constant volume/changing shape profiles or changing volume/changing shape profiles.

The pulforming process is similar to pultrusion in that primary reinforcing fibers are drawn off the supply rack and through a resin impregnating tank, and then pulled through a forming/curing die.

Straight Pulforming (Figure 2-26)

The straight pulforming process came into being as a means of making composite hammer handles, a changing volume/changing shape consumer product. Volume is changed by the following means: a bulk molding compound (BMC), purchased as an extruded rope on a supply spool, is fed through a monitoring device that measures a specific length to provide exact desired volume. The length is simultaneously cut and crimped around a single traveling roving strand at uniformly spaced intervals. The result resembles a string of beads proceeding downstream. A blivet press traveling at line speed forms the BMC pieces into

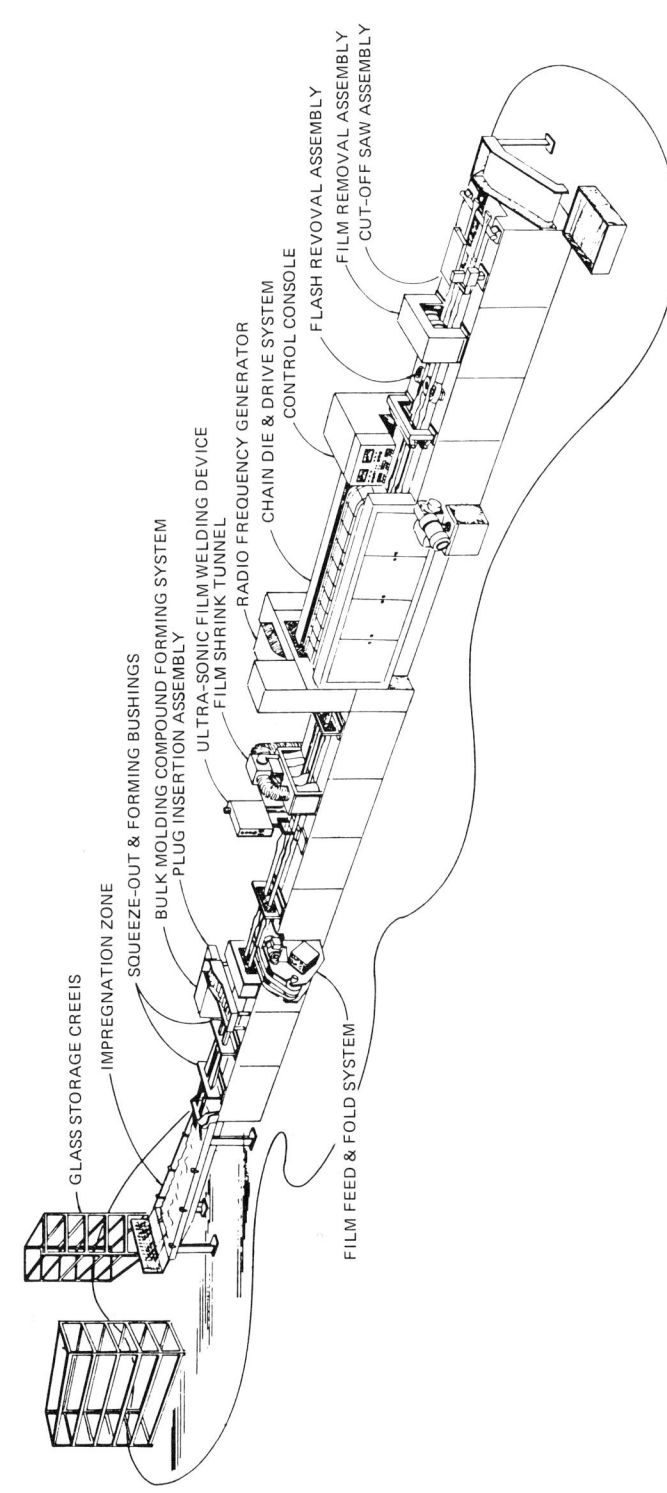

Figure 2-26. Straight pulforming equipment for manufacture of composite hammer handles.

64 ADVANCED THERMOSET COMPOSITES

the exact shape of the changing volume. Further downstream, all the primary reinforcing fibers come together to encapsulate that traveling center fiber with the formed BMC. Shrink film is then brought up from the bottom, convolutely folded around the package, and continuously, ultrasonically welded around it, creating a "sausage." The sausage passes through a shrink tunnel, which shrinks the film down very tightly, producing a very controlled package.

The package next enters a split female die, which is closed with a C-press as it travels at line speed downstream; the C-press injects the die into a belt clamping system that butts a whole stream of the dies together. The dies are heated as they traverse the clamping area; when they reach the opposite end, the die opens, releasing cured stock, and returns to the upstream, C-press end of the die section so that a continuous operation is effected. The part is then fed into a flying cutoff saw that cuts it to length.

Curved Pulforming

Constant volume/non–constant shape curved pulforming equipment is illustrated by Figure 2-27. A completely automatic piece of equipment, the machine

Figure 2-27. Pulforming equipment for manufacture of automotive leaf springs and other curved composite products.

pictured was developed specifically for making automotive leaf springs, although that is by no means an exclusive product market for this process. Again, reinforcing materials come in from supply racks, pass through an impregnating tank, are preheated in a radio frequency generator, and enter the die cavity, which is at the convergence of a stainless steel belt and the rotating die. The die cavity is in the shape of the part to be produced. The stainless steel belt clamps to the die face and runs with it continuously, thereby closing the fourth side of the die. In the length of that stainless steel belt, the part is finish-cured. It is then continuously peeled out of the die cavity and guided into the path of a flying cutoff saw. Existing machines are controlled by standard switch and relay logic; next generation machines, however, will utilize programmable controllers. Production rate, depending on specific spring configuration, is approximately two springs per minute. Not only is composite spring cost competitive, it is dramatically lower than the cost of the steel spring it replaces.

TAPE PLACEMENT

Of all the generic processes, tape placement most nearly simulates the hand techniques it seeks to automate. Reduced to its simplest common denominator, the concept is basically to utilize a machine, rather than a person, to lay strips of preimpregnated composite tape contiguously–but not overlapping—onto or into a flat or curved mold, and debulk and edge-trim them to a perfect fit. Since the tape itself has no integrity but is merely a limp, tacky ribbon of material, it is supplied by the manufacturer with backing paper that must be removed at the point of laydown. Tapered sections are accomplished by stepping-back succeeding layers until desired taper and buildup are attained. Each layer requires debulking; the resultant product is subjected to heat and pressure in order to achieve a final cure.

Tape placement machine evolution originated in the mid-fifties when the Air Force contracted with General Dynamics for development of a means of using the fiberglass 3 inch tapes available at the time to fabricate aircraft components. No specific component production was designated—the task assigned was a general one. General Dynamics contracted with Conrac, a southern California firm, who proceeded to build the first tape placement machine. The Conrac machine has been updated over the years and is still in service at General Dynamics.

Six-axis Machine Designed to Wrap Helicopter Blades

The next evolutionary step occurred some ten years later when process and equipment were developed for laying up by automated means a helicopter rotor

Figure 2-28. Six-axis automated tape lay-up system (ATLAS) for helicopter rotor blades.

blade. The result was the six-axis machine in Figure 2-28, christened ATLAS (Automated Tape Lay-up System).

The machine is designed so that rotor blade structures to be wrapped with tape are mounted between a headstock (which can rotate in the D-axis at from 1 to 10 rpm) and a tailstock, and are supported by two steady rests. The steady rests rotate with the blade and move in unison with the tape dispensing mechanism on the gantry (Figure 2-29). The gantry-mounted tape dispenser moves in the longitudinal X-axis, vertical Z-axis, traverse Y-axis, rotational C-axis, and $\pm 45°$ tilt A-axis. To facilitate high traverse rates, the moving gantry is a low inertia, one-piece composite structure.

The machine lays adjacent tape paths, automatically slitting 3 inch wide tape into narrow ribbons where extreme compound curvatures occur. Backing paper is taken up on a reel prior to tape laydown. Tapes are automatically debulked by the laydown/placement roller and automatically cut at the end of each run. The unit is programmed for stepping back of each layer to achieve correct thinning of the rotor skin as the trailing edge and blade tip are reached.

The ATLAS I utilized minicomputer numerical control, an advanced concept

FABRICATION TECHNIQUES 67

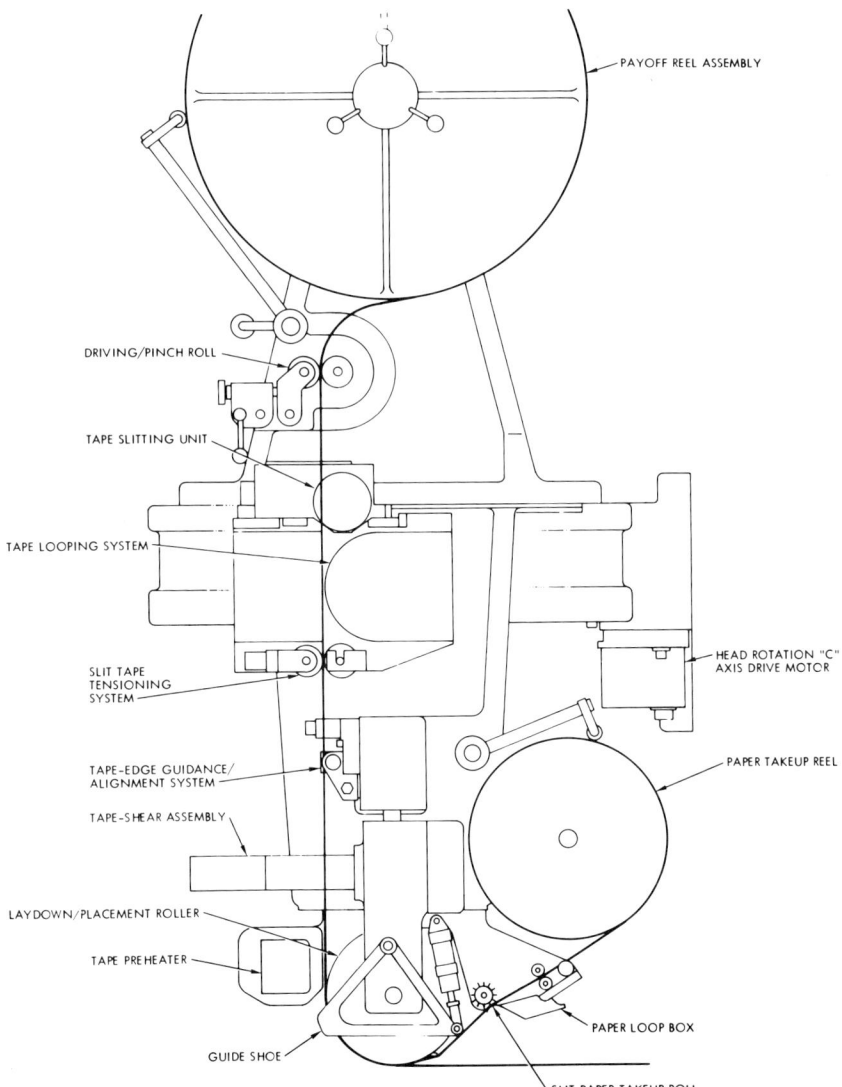

Figure 2-29. Tape placement head used on ATLAS and other early tape placement machine models.

at the time it was built. To simplify punched tape programming, the 3 inch tape laydown applicator was replaced with an optical line follower, permitting digitizing of shapes too complex to easily program mathematically.

Production Equipment

The most recent tape laying equipment innovation came about not as a result of a new application, but in response to the demand for true production equipment. Heretofore, all automated tape placement equipment has been essentially for prototyping and testing purposes. Now, however, composite components—in particular, advanced composite components (graphite, boron, aramid, and so forth)—have well proved their worth, and virtually all military aircraft manufacturers today specify composites for a large and diverse number of structural and nonstructural parts. All of these must be turned out on production equipment if this is to continue to be a viable and competitive manufacturing industry.

The primary drawback to sustained, high volume productivity in existing equipment is the incomplete shearing tendency of tape cutting devices. Even though reliability percentages as high as 93% have been achieved, the remaining 7% failure obviates the entire production system for the reason that when the tape head begins its laydown path following the cut cycle, if incomplete shearing has occurred, the head will bring the uncut, previously laid tape with it—and since the tacky prepreg will have been rolled down onto the preceding layer, that layer lifts too, virtually destroying the entire lay-up.

Two-phase tape preparation/tape laying equipment has been designed to overcome this and other less devasting problems of previous equipment. Phase 1 (Figure 2-30) measures and cuts the tape into predetermined lengths, simultaneously editing out undesirable sections, and winds the cut tape and backing paper onto casette reels (thus the name ACCESS—advanced composite cassette edit/shear system). The prepared cassette is then loaded onto phase 2, the ATLAS II tape placement machine (Figure 2-31). Directed by computer programmable controls, the five-axis machine lays the precut, edited tape onto the tooling bed at high production speeds. Function and operation of the tape head are otherwise essentially identical to the ATLAS I. In addition to positive, reliable tape cutting, a number of other advantages are incurred by this system, such as total versatility in cut angles; ability to cut on the fly, without stop-cut-accelerate downtime during tape laying; control of long, tapering tails on low angle cuts; ability to enable very small tape pieces to be cut and laid without the necessity for any handwork; and the ability to prestock cassettes prepared for specific part lay-up.

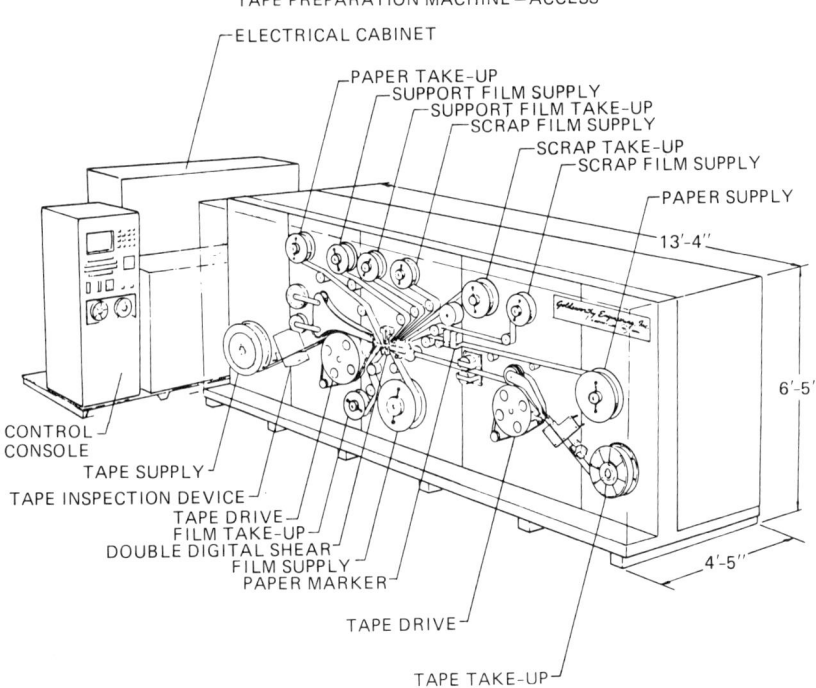

Figure 2-30. Two-phase tape placement system—ACCESS unit.

SUMMARY

Fabrication techniques for RP/C materials have evolved in the latter half of this century from pure hand lay-up by skilled craftsmen, through entrepreneurial adaptation of conventional metal working tools, into a high technology industry utilizing computer integrated manufacturing concepts for volume production of a large number of diverse consumer, industrial, and military end products. Applications for RP/C products have driven design efforts, and innovative designs have in turn stimulated new product applications. In addition to a considerable number of specialty machines that have been developed for a dedicated task, four basic, generic types of composite production equipment have emerged: filament winding, pultrusion, pulforming, and tape placement. Products from these materials and machines are found today in virtually every industry and activity, including sports and recreation, electrical components, furniture, building materials, pipe and tank manufacture, transportation, aeronautics and space ventures, and so forth. Potential applications for composite products ap-

70 ADVANCED THERMOSET COMPOSITES

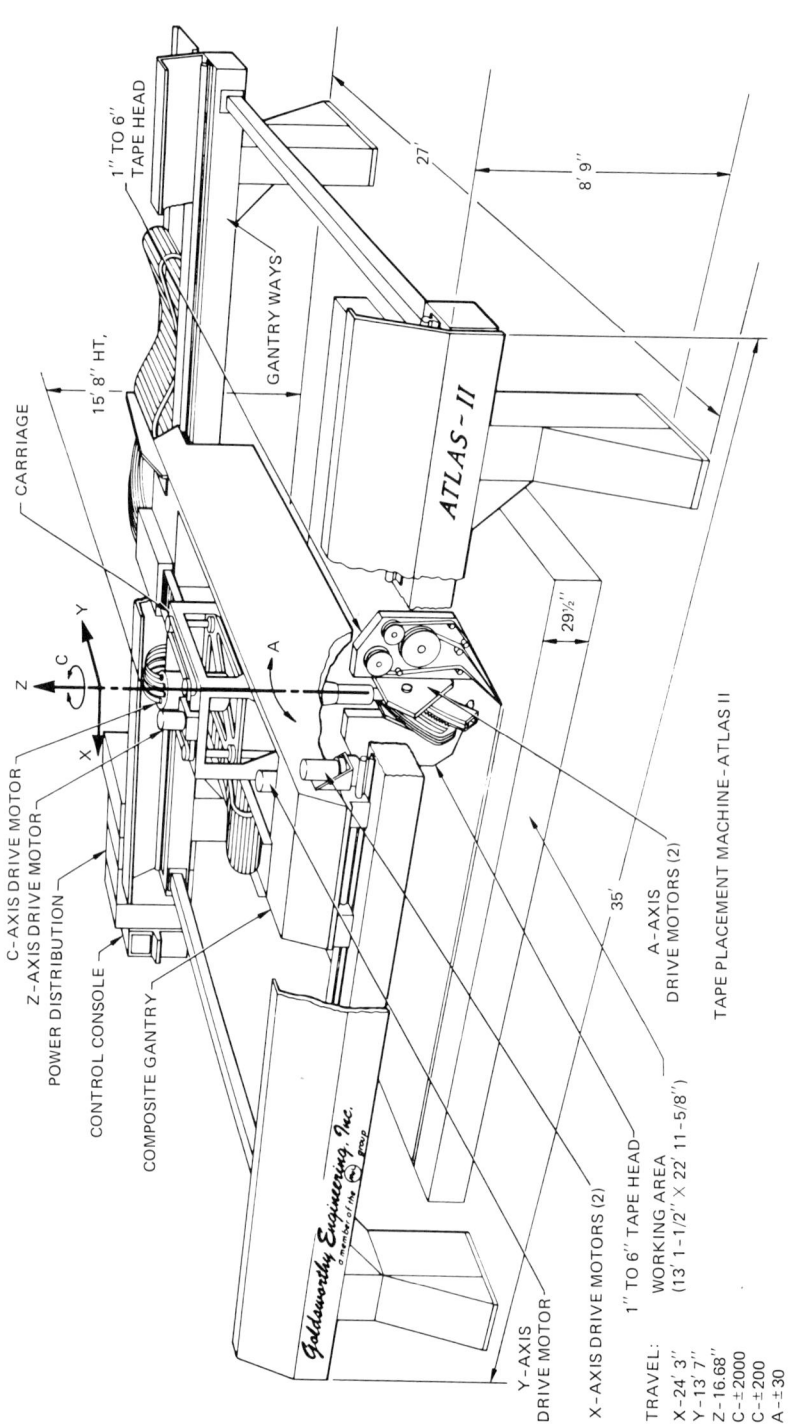

Figure 2-31. Two-phase tape placement system—ATLAS II unit.

pear to be nearly limitless. Continuing advancements in materials and processing will ensure that fabrication techniques will keep pace with future product requirements.

References

1. Hardesty, E. E., "Forming Fiber Reinforced Products," *Automation* (September 1970).
2. Jones, B. H., *Analysis for an Energy Absorbing Pultruded Beam* (November 1973). Final report of a contract carried out for Ford Motor Co. by the Composite Products Technology Center of Goldsworthy Engineering, Inc.
3. U.S. Patent 2,871,911 (Feb. 3, 1959), W. B. Goldsworthy (to Glastrusions, Inc.). Extrusion machine.
4. U.S. Patent 2,990,091 (June 27, 1961), W. B. Goldsworthy (to Glastrusions, Inc.). Pulling device.
5. U.S. Patent 3,556,888 (Jan. 19, 1971), W. B. Goldsworthy (to Glastrusions, Inc.). Pultrusion machine and method.
6. U.S. Patent 3,684,622 (Aug. 15, 1972), W. B. Goldsworthy (to Glastrusions, Inc.). Pultrusion machine (divisional).
7. U.S. Patent 3,674,601 (July 4, 1972), W. B. Goldsworthy (to Glastrusions, Inc.). Augmented curing of RP stock.
8. U.S. Patent 3,793,108 (Feb. 19, 1974), W. B. Goldsworthy (to Glastrusions, Inc.). Augmented cure (method).
9. U.S. Patent 3,960,629 (June 1, 1976), W. B. Goldsworthy (to Glastrusions, Inc.). Inductive heat curing of conductive fiber stock.
10. U.S. Patent 3,873,399 (March 25, 1975), W. B. Goldsworthy (to Goldsworthy Engineering, Inc.). Circular pultrusion apparatus.
11. U.S. Patent 3,701,489 (Oct. 31, 1972); 3,740,285 (June 19, 1973); and 3,738,637 (June 12, 1973); W. B. Goldsworthy (to Goldsworthy Engineering, Inc.). Box winding machine.
12. Hardesty, E. E., *Advanced Composite, High-modulus Tape Placement Machines* (April 1969). Presented to SAMPE 15th National Conference, Los Angeles.
13. Goldsworthy, W. B., "Thermoplastic Composites: The New Structurals," *Plastic World* (August 1984).
14. Hardesty, E. E., "Design and Construction of a Large, Fully Automated Tape Placement Machine for Aircraft Structures," *Composites* (November 1972).
15. U.S. Patent 4,462,946 (July 31, 1984), W. B. Goldsworthy (to Goldsworthy Engineering, Inc.). Straight pulforming, changing volume/changing shape.
16. U.S. Patents 4,440,593 (April 3, 1984) and 4,469,541 (Sept. 4, 1984), W. B. Goldsworthy (to Goldsworthy Engineering, Inc.). Curved pulforming, changing shape and/or volume.
17. U.S. Trademark "Pulformer" 1,187,389 (Jan. 26, 1982), W. B. Goldsworthy (to Goldsworthy Engineering, Inc.).
18. U.S. Patent 3,801,407 (April 2, 1974), W. B Goldsworthy (to Goldsworthy Engi-

neering, Inc.). Apparatus and method for producing plastic reinforced sheet laminates.
19. U.S. Patent 4,402,778 (Sept. 6, 1983), W. B. Goldsworthy (to Goldsworthy Engineering, Inc.). Method for producing FRP sheet structures.
20. U.S. Patent 4,420,359 (Dec. 13, 1983), W. B. Goldsworthy (to Goldsworthy Engineering, Inc.). Same as above.
21. U.S. Patent 4,125,423 (Nov. 14, 1978), W. B. Goldsworthy (to Goldsworthy Engineering, Inc.). RP tapered rod products and method, and apparatus for producing same.
22. U.S. Patent 4,032,383 (June 28, 1977), W. B. Goldsworthy and H. E. Karlson (to McDonnell Douglas). Continuous 3D foaming machine and method.
23. U.S. Patent 3,966,533 (June 29, 1976), W. B. Goldsworthy (to Goldsworthy Engineering, Inc.). On-site wall structure formation.
24. Anon.,"Energy Content of Reinforced Plastics Materials," *International Reinforced Plastics Industry (IRPI)* (November 1981).

Reading List

Air Force System Command, Wright-Patterson Air Force Base, Ohio (Structures Div., Air Force Flight Dynamics Laboratory), *Advanced Composites Design Guide*, vol. III, IV, and V (3rd ed. rev.), January 1977. Prepared under Contract No. F33615-74-C-5075 by Los Angeles Aircraft Div. of Rockwell International.

American Astronautical Society Proceedings, "Space Manufacturing 1983," *Advances in the Astronautical Sciences*, **53**, ASA Publications, San Diego (1983).

Army Aviation System Command, St. Louis, *Development of an Automated Layup System for Monofilament Fiber Structures*. Final report prepared under Contract DAAJ01-70-C-0049(PIG) by Boeing Vertol Co.

Ashton, J. E. and J. M. Whitney, *Theory of Anisotropic Plates*, Technomic, 1970.

Delmonte, J. *Technology of Carbon and Graphite Fiber Composites*. New York: Van Nostrand Reinhold, 1981.

Goldsworthy, W. B., *Volume Component Manufacturing*. Short course presented to UCLA Extension, Composite Materials, Sept. 17–21, 1984.

International Reinforced Plastics Industry, a journal published in Great Britain.

Johnson, J. E., J. W. Lee, and R. M. DuPuis, *Reinforced Plastic Beam Shapes*. Presented to ASCE Conference on Selection, Design and Fabrication of Composite Structures, November 1972.

Jones, B. H. *Probablistic Design and Reliability*, New York: Academic Press, 1974.

Katz, H. S. and J. V. Milewski, Eds., *Handbook of Fillers and Reinforcements for Plastics*, Van Nostrand Reinhold, New York, 1978.

Lubin, G., Ed., *Handbook of Composites*, New York: Van Nostrand Reinhold, 1982.

Materials Evaluation, **43**, 1084 (1984).

Morgan, P., Ed., *Glass Reinforced Plastics*, Iliffe & Sons, London, 1958.

Roark, R. J., *Formulas for Stress and Strain*, McGraw-Hill, New York, 1965.

SAMPE Journal, **20** (1984).

SAMPE Proceedings, *2020 Vision in Materials for 2000*, Vol. 15. SAMPE, Azusa, Calif., 1983.
SAMPE Quarterly, **15** (1983).
SAMPE Quarterly, **15** (1984).
Schwartz, M. M., *Composite Materials Handbook*, McGraw-Hill, New York, 1984.
Stone, K. L., *SAMPE Quarterly*, **15** (1984).
Tsai, S. W., Ed., *Journal of Composite Materials*, Technomic, Lancaster, Pa.
Vought Systems Div., *Improved Automated Tape Laying Machine*, Technical Report AFML-TR-73-307, February 1974.

3
PROPERTIES AND PERFORMANCE REQUIREMENTS

James C. Leslie

President
Advanced Composite Products and Technology, Inc.
Huntington Beach, California

PROPERTIES OF ADVANCED COMPOSITES

For the purpose of this chapter, *advanced composites* are defined as structures in which the organic matrices are reinforced by continuous, oriented, high strength and/or high modulus fibers. Fiber reinforced metal matrix systems are also included in the general category of advanced composite materials but are not discussed here.

The data presented herein are, to the best of the author's ability, accurate, referenced, and/or traceable to reputable sources; however, the reader is cautioned to use the numbers given only for initial screening evaluations. In all cases specific test or manufacturer's data should be obtained for final component design/analysis. Said specific data should be for the actual materials (fiber, matrix, sizings, etc.) to be used and should apply to the static, dynamic, and environmental conditions under which the structural unit is to operate.

Information is presented on all the major reinforcement fibers. These are the properties which, when translated through the composite, provide unique structures. Room temperature laminate data are shown for typical aerospace grade epoxy (and polyester) matrices. Sufficient information is provided to allow first design approximations and reinforcement trade-off analysis. Space does not allow discussion of the many matrix systems herein. Having selected a basic reinforcement system, the reader should then either contact current hardware or materials manufacturers and consultants, or review references 1, 2, or 3. The data presented in this chapter are those which have been most helpful to the author and which, in his experience, are most requested.

GLASS REINFORCED COMPOSITES

Fiberglass is the reinforcement most often used in fabricating structural composite hardware. The composites industry effectively started with the introduction of commercial glass fibers by Owens-Corning in 1939. Since then, glass has been investigated as a reinforcement for most organic matrices and in forms from nonoriented short fibers to continuous, precisely aligned tows. As glass demonstrated the advantages of high strength reinforced composites, the industry responded by developing other high strength/high modulus fibrous materials. As shown in Table 3-1, glass fibers still hold about 90% of the reinforced plastic structures market.

Glass Fibers

Many special purpose glass fibers have been developed (see reference 2). Only E (electrical grade), S (high strength), and the "quartz" varieties will be described in detail here. These are the three glasses most often used in advanced composites.

E glass is by far the most used of the three. It is the lowest priced, under a dollar, per pound; provides a chemically and environmentally stable, high strength reinforcement; and offers very attractive thermal and electrical properties.

Commerical S glass costs three to four times as much as E. The current, second generation, S glass was introduced in the late 1960s and effectively replaced the more expensive form developed in the early sixties for high performance aerospace applications. The major advantages of S glass are 35 to 40% higher tensile strength, 20 to 25% higher compressive strength, 2 to 4% lower density, and 18 to 20% higher modulus. S glass also has higher resistance to attack in strong acids and, correspondingly, is degraded faster by strong bases than E glass.

Table 3-1. Fiber Reinforcement Sales [4].

Fiber Type	U.S. Consumption (metric tons)		
	1983	1984	1985
Glass	300,000	345,000	350,000
Aramid	1,200	1,600	2,000
Carbon/graphite	1,000	1,400	1,800
Other	5	10	15

Table 3-2. Properties of Glass Fibers.

Property	E Glass	S Glass	Quartz Glass*
Mechanical/Physical			
Tensile, ultimate, ksi/MPA	500/3450 [6]	665/4580 [7]	130/896 [2]
Tensile, modulus, Msi/GPA	10.6/72.4 [6]	12.6/86.8 [7]	10/6.89 [8]
Tensile, ultimate elongation @ 72°F	4.8 [7]	5.4 [7]	0.17 [8]
Poisson's ratio	0.22 [6]		
Density, $(lb/in.^3)/(g/cm^3)$	0.94/2.60 [7]	.090/2.49 [7]	
Index of refraction, 32°C	1.549 [6]	1.523 [5]	1.4585 [8]
Ultraviolet transmission	Opaque		
Acoustical, velocity of sound, ft/sec			
Calculated	17,500 [5]	19,200 [5]	
Measured	18,000 [5]		
Thermal			
Coefficient of thermal expansion (10^{-6}), (in./in.-°F)/(cm/cm-°C)	2.8/1.57 [7]	3.1/1.74 [7]	0.3/0.54 [8]
Coefficient of thermal conductivity, btu-in./hr/ft^2/°F	7.2 [6]		
Specific heat 75°F, cal/g-°C	0.192 [9]	0.176 [9]	0.23 [8]
Softening point, °F/°C	1540/838 [6]	1778/970 [9]	3038/1670 [8]
Electrical			
Dielectric constant @ 72°F			
6 Hz	5.9–6.4 [2]	5.0–5.4 [2]	
1 MHz	6.33 [9]	5.34 [9]	3.7 [8]
10 MHz	6.13 [9]	5.21 [9]	(MHz not specified)
Power factor (loss tangent) @ 73°F			
1 MHz	0.001 [9]	0.002 [9]	0.0002 [8]
10 MHz	0.0039 [9]	0.0068 [9]	

*99.5% SiO_2 (Astroquartz) [8].

Structural quartz fibers cost on the order of 100 times more than E glass fibers. They are finding extensive use, however, in cryogenic, electrical, high temperature, and other areas which require the special properties supplied by very pure quartz.

Table 3-2 is a comparison of the major physical, electrical, thermal, and mechanical properties of E, S, and quartz glass fibers. The high tensile strength capabilities—500,000 and 665,000 of E and S glass, respectively—have been the major driving force in their use. The 12,600,000 modulus of S glass, combined with its good translation of composite compressive strength, also brings it into contention with Kevlar ®* in many applications. The extremely low con-

---

*Registered trademark of E. I. du Pont de Nemours & Company.

Table 3-3. Glass Yarn Nomenclature [10].

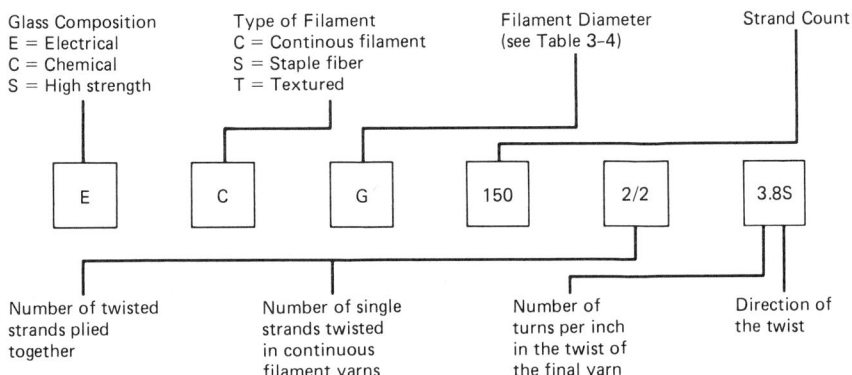

ductivity of glass has resulted in the widespread use of fiberglass composites as electrical hardware components. Similarly, glass and Kevlar are unique in their combined structural/transmission capability for radar and electronic packaging applications. It is this type application where quartz glass also provides absolutely unique characteristics.

Available Forms of Glass Reinforcement

The widespread use of glass fibers has resulted in commercial availability in almost every conceivable form (in fact, the suppliers have stated that given a "sufficient market" they will modify their product to meet any attainable demand).

The fiber suppliers provide continuous yarns or tows. See Tables 3-3 and 3-4 for a quick summary of typical strand sizings and tow designations. Reference 5 provides in-depth analysis of the glass fiber yarn/tow identification and characteristics. Glass is also supplied in chopped, mat, or paper forms and as molding compounds, with many possible fiber lengths and various degrees of orientation possible. Detailed identification is beyond the scope of this chapter. For more information, consult reference 2 and/or the major fiber suppliers.

Glass reinforced advanced composite structures often employ filament winding, "layed-up" unidirectional prepeg, or glass cloth. Table 3-5 is a listing of some typical structural glass fabrics. As shown, fabrics are available in many weaves, thicknesses, and weights. Again, if the exact product you want isn't available, many of the weavers will contract for experimental quantities, and most certainly will set up and weave (or knit) large quantity orders. For more data on available fabrics, contact individual suppliers.

Table 3-4. Glass Filament Diameters [5].

Filament Designation		Nominal Diameter Inches
U.S.	Metric	
B	3.5	.00015
C	4.5	.00019
D	5	.00021
DE	6	.00025
E	7	.00029
G	9	.00036
H	10	.00043
K	13	.00053

Table 3-5. Representative Glass Fabrics [12].

				Construction				
				Count		Yarn*		
Style	Weave	Wt (oz/yd^2)	Thickness (in. × 10^{-3})	Warp		Fill	Warp	Fill
			Bidirectional					
104	Plain	0.58	1.2	60	×	52	900 1/0	1800 1/0
108	Plain	1.43	2.0	60	×	47	900 1/2	900 1/2
116	Plain	3.16	3.5	60	×	58	450 1/2	450 1/2
120	Crowfoot satin	3.16	4.0	60	×	58	450 1/2	450 1/2
128	Plain	6.0	7.0	42	×	32	225 1/3	225 1/3
143	Crowfoot satin	8.9	9.0	49	×	30	225 2/3	450 1/2
181[†]	8 Harness satin	8.9	8.5	57	×	54	225 1/3	225 1/3
7781[†]	8 Harness satin	9.5	9.0	57	×	54	75 1/0	75 1/0
1543 [13]	Crowfoot satin	9.4	8.6	49	×	30	150 2/2	450 1/2
7500	Plain	9.75	14	16	×	14	75 2/2	75 2/2
184	8 Harness satin	25.9	27	42	×	36	225 4/3	225 4/3
1597 [14]	12 Harness satin	40	45	30	×	30	150 4/4	150 4/4
			Unidirectional					
143	Crowfoot satin	8.9	9	49	×	30	225 2/3	450 1/2
341	Crowfoot satin	8.8	9.7	30	×	49	450 1/2	225 3/2
1543	Crowfoot satin	8.8	9	49	×	30	150 2/2	225 1/0
1667	Plain	3	4.3	60	×	12	150 1/0	150 1/0

*All fibers are ECD; see Tables 3-2 and 3-3.
[†]7781 is the now accepted "commercial" version of the widely used "mil spec" 181 style fabric.

Glass Composite Properties

It would take many volumes to publish all of the data on glass reinforced plastics. Table 3-6 presents a comparison of the properties possible with some of the common methods of manufacturing. This information is limited to one matrix (polyester) but illustrates the range of mechanical, physical, and electrical properties. In general, injection molding with thermoplastic uses short fibers and is limited to a 30+% fiber weight loading. Bulk molding with longer fibers allows higher fiber loadings and, therefore, can achieve higher mechanical properties. By definition, advanced composites include only continuous and oriented fibers molded in such a way as to approach maximum mechanical property values. Certainly the properties attainable by pressure curing lay-ups, pultrusion, and filament winding fit the definition. Sheet molding compound and the oriented sheet molding techniques such as XMC®* should also be included. For in-depth information on injection and compression molded composites, see references 11 and 2.

The mechanical properties shown for "filament winding" in Table 3-6 are also representative (possibly low) of what can be achieved by pressure curing of "layed-up" unidirectional and cloth laminates. They are generally applicable for most modern matrix systems.

Glass Laminate Properties

Typical unidirectional, 0°, glass/epoxy laminate properties are presented in Table 3-7. These values are those used in-house at Advanced Composite Products and Technology, Inc., as first approximations for composite hardware design. They represent attainable room temperature properties for a 60% fiber volume laminate with a wet-winding (or pressure cured lay-up) epoxy. Substantially different properties are possible with different matrices, after exposure to environmental or vibrational degradation, and with different fiber and fiber sizings. Most properties are also highly temperature dependent. While listings such as Table 3-7 are useful for first design approximations, the reader is cautioned to obtain specific data for the exact system and application being considered. Computation procedures and programs are available for final design/analysis.

Figure 3-1 illustrates the decrease in mechanical strength as off-axis loads are applied. It is a plot of 0°, axial/tensile strength as function of the helical, ±, winding angle. Again, the data plotted herein represent "first approxima-

---

*Registered trademark of PPG Industries Inc.

Table 3-6. Typical Properties, Fiberglass Composites: Different Methods of Fabrication [6].

Polyester (Styrene Monomer)	ASTM Method	Open Molding		Matched Die Compression Molding				Filament Winding	Pultrusion (Rod Stock)
		Lay-up	Spray-up	Bulk Mldg Cpd	Sheet Mldg Cpd	XMC Composite	Injection Molding RTP		
Mechanical Properties									
Tensile strength,									
psi × 10^3 [a,b]	D638	9–50	5–18	3–10	8–20	70–90	6–30	80–200[a]	70–180
MPa		62–345	35–124	21–69	55–138	480–620	41–207	550–1380[a]	480–1240
Tensile modulus,									
psi × 10^6	D638	0.6–4.5	0.8–1.8	1–2.5	1–2.5	6.0–6.5	0.5–1.8	4–8	4–6
GPa		4–31	6–12	7–17	7–17	41–45	3–12	28–55	28–41
Compressive strength, (edgewise)									
psi × 10^3	D695	18–50	15–30	15–30	15–30	79/28	6–26	50–80	40–100
MPa		124–345	103–207	103–207	103–207	545/193	41–179	345–550	276–690
Shear strength,									
psi × 10^3	c	4–6				3.5–9		7–10	5–10
MPa		28–41				24–62		48–69	35–69
Impact strength, notched Izod,									
ft lb/in.	256	5–30	5–15	1.5–10	8–22	65	1–5	40–60	
J/m		270–1600	270–800	80–535	427–1170	3470	54–270	2135–3200	

Physical Properties

Property	ASTM								
Density, specific gravity[d] lb/in.³	D792	1.4-2.1 0.051-0.076	1.4-1.6 0.051-0.058	1.6-2.3 0.058-0.083	1.6-2.6 0.058-0.094	2.0 0.073	1.1-1.7 0.040-0.061	1.7-2.2 0.061-0.079	1.7-2.1 0.061-0.076

Electrical Properties

Property	ASTM								
Dielectric constant,	D150								
60 cps		3.8-6.0	3.7-6.0	5.3-7.3	4.4-6.3		2.4-4.2[e]	4.2-5.3	4.0-6.0
10³ cps		4.0-6.0	4.0-6.0	4.0-6.8	4.4-6.1		2.4-4.0		4.0-6.0
10⁶ cps		3.5-5.6	3.6-6.0	5.2-6.4	4.2-6.0		2.4-3.9	4.0-5.2	4.0-6.0
Dissipation factor,	D150								
60 cps		0.01-0.05	0.01-0.04	0.01-0.2	0.007-0.2		0.001-0.03		
10³ cps		0.01-0.06	0.01-0.05	0.01-0.2	0.007-0.2		0.001-0.025	0.018-0.05	
10⁶ cps		0.01-0.03	0.01-0.03	0.01-0.02	0.01-0.02		0.0015-0.26		
Dielectric strength, V/mil (short time)	D149	300-600	200-450	300-450	300-450		400-600	200-400	300-350[f]

[a] For XMC composite values are parallel to continuous reinforcement, except compressive strength for which directional properties of XMC-3 (parallel/transverse) are given. XMC-3 provides 9-15 × 10³ psi and 1.4-1.8 × 10⁶ psi transverse tensile strength. For filament wound structures, values are by ASTM D2290 and D2343 for filament and strand tensile strengths, respectively.
[b] Ultimate tensile elongation ranges from 0.3 to 2.7% for FGRTP thermoset polyesters, and from 1 to 5% for FGRTP thermoplastics.
[c] ASTM D3846, twin notch, in-plane shear in compression, except values for filament winding which are apparent horizontal shear, ASTM D2344, by the short beam method.
[d] May be calculated as follows: laminate sp. gr. equals the reciprocal of the sums of the ratios of weight fraction to sp. gr. for each component in the laminate.
[e] Dielectric constants and dissipation factors are substantially higher for nylons conditioned to 50% RH.
[f] Range applies to sheet. Parallel dielectric strength for rods is 50 KV/in.

Table 3-7. Typical 0° E Glass/Epoxy Laminate Data.*

Property	E Glass	S Glass
Tensile strength, ksi/MPa	165/1137	195/1344
Tensile modulus, Msi/GPa	6.5/45	7.7/53
Compressive strength, ksi/MPa	111/765	129/889
Shear strength, ksi/MPa	6.1/42	6.1/42
Shear modulus, Msi/GPa	.28/1.9	.28/1.9
Poisson's ratio,		
0°	.272	.272
90°	.092	.08
Density, (lb/in.^{3})/(g/cm^3)	.072/1.99	.071/1.97

*In-house "first approximation" data as used at Advanced Composite Products and Technology, Inc. for a 60% fiber volume in an epoxy matrix.

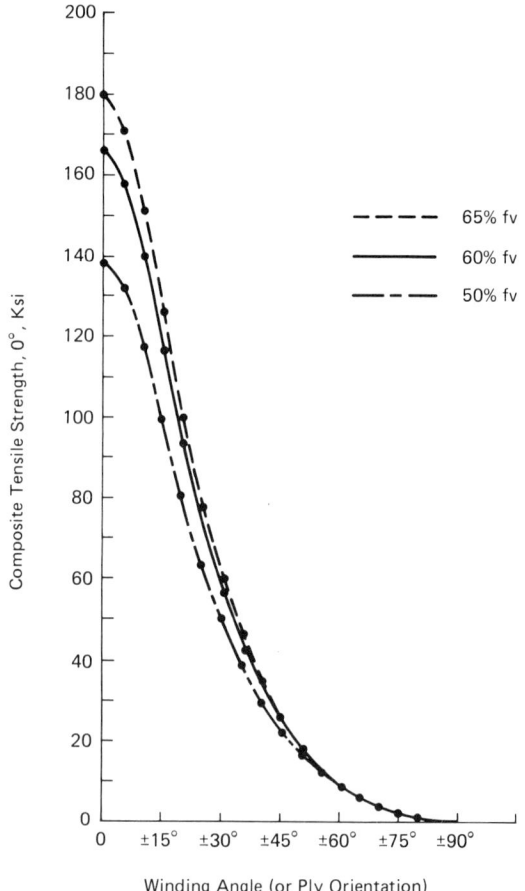

Figure 3-1. Composite tensile, E glass/epoxy, varies with winding angle and fv.*

---

*"In-house, first approximation data," Advanced Composite Products and Technology, Inc.

Table 3-8 E Glass Cloth/Epoxy Laminates [15].

Property	100% Glass	100% Kevlar	100% Carbon	50% Carbon 50% Glass
0° Tensile strength, ksi	72	82	82	56
0° Tensile modulus, Msi	4.3	6.1	10.3	7.5
0° Flexural strength, ksi	97	69	87	103
0° Flexural modulus, Msi	3.8	4.8	8.9	7.3
Short beam shear, ksi	7.5	4.0	8.5	9.5
Cost/yd, $	5	18	55	30

tion'' calculations as used in-house at Advanced Composite Products and Technology, Inc. Note that the contribution of the resin matrix is assumed to be 0, thus providing a degree of conservatism to the resultant off-axis determination. Three lines are shown, illustrating the effect of fiber loading, fv. The lower 50% fv is generally accepted as a lowest desired level, while the 60 to 65% lines represent nominal goals. Similar curves can be generated, by computer, for different fibers, resins, temperatures, and loading conditions. Input to the programs requires data such as those given in Table 3-7 for the specific materials and applications being considered.

Table 3-8, reproduced from reference 15, compares mechanical properties in laminate made from woven cloth. It is comparable to Table 3-7 and also shows data for Kevlar, carbon/graphite (standard fiber), and glass/carbon hybrid cloth. Note that in using cloth in laminates, similar properties are attained in both the 0° and 90° directions, and 0°/90° lay-up of unidirectional fibers would provide directly comparable numbers.

KEVLAR REINFORCED COMPOSITES

Kevlar is the registered trademark for one group of E. I. du Pont's family of man-made, aromatic polyamide fibers. These fibers are characterized by low density, high tensile strength, medium modulus, and extreme toughness. They are available in chopped, mat, woven, yarn, and tow forms. Kevlar is finding wide acceptance in aerospace structures, pressure vessels, tires, ballistic protection, sporting goods, marine, automotive, industrial, military, and electrical applications. Kevlar fibers are useful with many commonly employed organic composite matrices.

Kevlar Fiber

Kevlar reinforcement fibers were commercially introduced by du Pont in 1972. Their low density and high strength characteristics resulted in evaluation for reinforcing lightweight aerospace structures. As with other reinforcement materials, this work soon led to exploitation of other unique properties. To date, only two modifications, Kevlar 29 and Kevlar 49, have been widely investigated and used extensively in advanced composite structures. Two superior sources of information of Kevlar fibers and composite structures are the E. I. du Pont Company (especially its data manual [16]) and the chapter on Kevlar in Lubin's *Handbook of Composites* [2].

Kevlar29. Kevlar 29 was introduced as the commercial (as opposed to aerospace) version of Kevlar 49. It is widely used as tire cord, for ropes and cables, and in ballistic applications. Table 3-9 shows that the fibers are essentially equal in physical characteristics and tensile strength, varying mainly in stiffness (modulus). Lubin [2] also shows some minor differences in chemical resistance.

Kevlar 49 Fiber. Kevlar 49 is primarily recognized for its extremely high tensile strength:density ratio. It also finds use through its thermal, electrical, and chemical characteristics. Fiber properties and commercially available forms of major interest are summarized here. Typical composite properties are given in the following section.

Kelvar 49 is available as short length chopped fibers; in paper, mat, felt, or "nonwoven" blanket forms; as yarns or tows; or in woven fabrics. The yarns or tows available from du Pont are shown in Table 3-10. Fabrics are available in most "standard" bidirectional structural weaves and in special "unidirectional and hybrid forms. Table 3-11 is a compilation of typical fabrics.

Kevlar is extremely "tough." This fact is illustrated in Table 3-12 which shows the tensile and tear strength of several fabrics.

The thermal characteristics of Kevlar also leads to its use in applications requiring very low coefficients of thermal expansion and/or low thermal conductivity. Tables 3-13 through 3-15 summarize these properties. Note especially that in the fiber direction, a negative coefficient of expansion of $-1.1 \times$

Table 3-9 Comparison of Kevlar 29 and 49 Fibers.

Property	Kevlar 49		Kevlar 29	
Tensile strength, ksi/MPa	525/3620	[16]	525/3620	[17]
Tensile modulus, Msi/GPa	18/124	[16]	9/62	[17]
Tensile ultimate Elongation, %	2.5	[16]	4.0	[17]
Calculated axial compression modulus, Msi/GPa	11/75.8	[2]	5.8/40.7	[2]
Density, $(lb/in.^3)/(g/cm.^3)$	0.52/1.44	[16]	0.52/1.44	[17]
Filament diameter, in./cm	.00047/.00119	[16]	.00047/.00119	[17]
Filament cross section	Round		Round	

PROPERTIES AND PERFORMANCE REQUIREMENTS   85

Table 3-10. Yarns and Rovings of Kevlar 49 Aramid [16].

Denier	Decitex	Yield (yd/lb*)	Yield (m/kg)	Filaments	Twist
195	215	22,895	46,155	134	0
380	420	11,749	23,684	267	0
1140	1270	3,916	7,895	768	0
1420	1580	3,144	6,338	1000	0
4560	5100	980	1,973	3072	0
7100	7900	630	1,268	5000	0

*Based on nominal denier.

10^{-6} in./in./°F is exhibited at moderate temperatures (0 to 100°C) and that it becomes more negative (-2×10^{-6}) above 200°C. The heat of combustion for Kevlar is reported [16] as 14,986 Btu/lb (34.8 J/kg) which is directly comparable to typical resin systems (828/NMA/BDMA at 12,710 btu/lb [16]). Du Pont also presents [16] direct smoke generation and vertical flammability data for typical Kevlar fabrics. These are reproduced in Table 3-16.

Kevlar demonstrates good chemical and thermal stability. Table 3-17 is a summary of tensile strength retention data following exposure to common chemicals. Minimal degradation is shown for exposure to organics and weak reagents. Strong acids and bases attack the fiber. Table 3-18 indicates no problem with long-term use in air at temperature below 320°F (160°C).

Data are also presented by du Pont on two areas of possible concern with Kevlar fiber: moisture pickup and ultraviolet degradation. As shown by Figure 3-2 Kevlar fiber has a strong affinity for moisture. As a result, it is necessary to dry Kevlar and then control humidity exposure prior to, and during, impregnation with moisture sensitive resin systems. Table 3-19 shows that the discoloration experienced when Kevlar is exposed to ultraviolet is accompanied by minor strength loss. This loss would only be a factor in very thin structures which could not be protected by paint or some other opaque coating.

Kevlar 49 Composites

Tables 3-20 and 3-21 are summaries of the typical mechanical properties of unidirectional Kevlar 49/epoxy laminates. One of the outstanding features of Kevlar is its tensile strength. The tensile stress-strain relationship of unidirectional, 0°, epoxy laminates shows a straight line plot to over 200,000 psi at a little less than 2% elongation (strain-to-failure). This high tensile translation combined with the low (0.05 lb/in.^{3}) density provides a very high specific strength (200,000 psi $\div$ .05 lb./in.^{3} = 4×10^6 inches). This property makes

86 ADVANCED THERMOSET COMPOSITES

Table 3-11. Kevlar Fabrics [18, 19].

Fabric Designation	Weave	Construction-Tow/yarn Denier* Warp (length) × Fill (across)	Wt (oz/yd^2/g/m^2)	Thickness (in. × 10^{-3}/mm)
Bidirectional Fabrics				
120	Plain	34 × 34-195	1.8/61	4.5/1.14
220	Plain	22 × 22-380	2.2/75	4.5/1.14
31	Crowfoot	41 × 40-195	2.2/75	4.5/1.14
181	8 Harness satin	50 × 50-380	5.0/170	9/2.29
281	Plain	17 × 17-1140	5.0/170	10/2.54
285	Crowfoot	17 × 17-1140	5.0/170	10/2.54
328	Plain	17 × 17-1420	6.8/231	13/3.3
335	Crowfoot	17 × 17-1420	6.8/231	12/3
500	Plain	13 × 13-1420	5.0/170	11/2.8
Unidirectional				
143	Crowfoot	100 × 20-380 × 195	5.6/190	10/2.54
243	Crowfoot	38 × 18-1140 × 380	6.7/227	13/3.3
Woven Roving				
1033	8 × 8 basket	40 × 40-1420	15.0/509	26/3.3
1350	4 × 4 basket	26 × 22-2130	13.5/458	25/6.4

Hybrid Fabrics†

Carbon graphite/Kevlar				
F3C812	Plain	12.5-Celion 3000 × 1140 Kevlar	.55/18.7	9.71/2.47
F3T825	Crowfoot	24 × 24-Thornel 300, 3K × 195 Kevlar	6.36/216	8.19/2.08
F3T860	8 Harness satin	24 × 24 Warp & Fill: 50% T300, 3K 50% 1140 Kevlar	10.0/339 10.0/339	12.6/3.2 12.6/3.2
F4D928	5 Harness satin	17 × 17 Warp & Fill: 75% 1140 Kevlar 25% 75-1/0 Glass	4.5/153	5.6/1.4
F4D903		21 × 8 1140 Kevlar × 150-1/0 glass	3.3/112	5.0/1.3

*Denier = weight in grams of 9000 meters of yarn.
†"Nonstandard" weaves produced by Hexcel, Inc., Dublin, Calif., available by special order only.

Table 3-12. Tensile and Tear Strength of Fabrics of Kevlar 49 Aramid [16].

Fabric Designation	Weave	Tensile Strength (Warp/Fill Direction)		Tongue Tear Strength		Trap Tear Strength	
		(lb/in.)	(kN/m)	(lb)	(N)	(lb)	(N)
120	Plain	250/250	44/44	60/60	267/267	22/22	98/98
181	8 Harness satin	700/700	123/123	110/110	489/489	56/56	249/249
281	Plain	650/650	114/114	105/105	467/467	43/43	191/191
285	Crowfoot	650/650	114/114	—*	—	40/40	178/178
328	Plain	700/700	123/123	120/120	534/534	65/65	289/289
143	Crowfoot	1300/125	228/22	—*	—	15/70	67/311
243	Crowfoot	1500/300	263/53	—*	—	20/100	89/445

	Property	STM No.
Test Methods:	Weight	D1910
	Tensile strength	D1117
	Tongue tear	D2261
	Trapezoidal tear	D2263

*Construction too loose for testing.

Kevlar 49 the first choice for pure tensile critical space structures such as pressure vessels.

On the negative side is the poor translation of Kevlar in compression. Here the stress-strain curve, shown in Figure 3-3 shows a straight line relationship to only about 30,000 psi. Above this point, a ductile-like response begins which continues until, at slightly above 38,000 psi, the curve becomes a horizontal line where continued loading results only in an increase in the strain. Lagasse of du Pont in reference 18 states that this behavior is caused by an internal buckling of the Kevlar filaments. Lagasse continues by illustrating how the combined tensile/compression loading of flexure specimens also result in a non-

Table 3-13. Thermal Coefficient of Expansion of Kevlar 49 [16].

	Temperature Range		Thermal Coefficient	
	(°C)	(°F)	(cm/cm °F)	(in./in. °F)
Longitudinal direction	0–100	32–212	-2×10^{-6}	-1.1×10^{-6}
	100–200	212–392	-4×10^{-6}	-2.2×10^{-6}
	200–260	392–500	-5×10^{-6}	-2.8×10^{-6}
Radial direction	0–100	32–212	59×10^{-6}	33×10^{-6}

Table 3-14. Thermal Conductivity of Fabrics and Felts of Kevlar and Other Materials [16].

Material	Form	Basis Weight (oz/yd^2)	Basis Weight (g/m^2)	Thickness (in.)	Thickness (mm)	Lag Time (sec)	Conductivity (cal/cm^2 sec)	Temperature Rise (25 sec) (°F)	Temperature Rise (25 sec) (°C)
Kevlar	Fabric 1-ply	9.8	333	0.030	0.76	0	0.324	108	60
Kevlar	Fabric 3-ply	29.4	998	0.085	2.16	3	0.162	54	30
Kevlar	Nonwoven felt	27.0	917	0.105	2.67	1.5	0.084	28	16
Fiberglass	Fabric 1-ply	8.4	285	0.012	0.30	0	0.600	200	111
Fiberglass	Fabric 8-ply	67.2	2282	0.085	2.16	5.1	0.105	35	19
Asbestos	Fabric	40.8	1386	0.090	2.29	2.5	0.168	55	31

TEST APPARATUS:

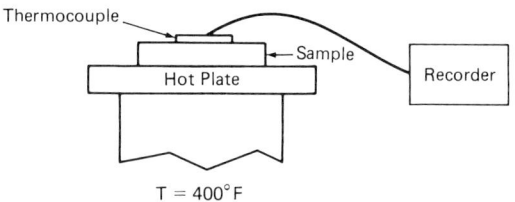

T = 400°F

LAG TIME IS TIME BETWEEN SAMPLE IN CONTACT WITH HOT PLATE AND ANY PERCEPTIBLE RECORDER READOUT.

E. I. Du Pont de Nemours & Company.

Table 3-15. Specific Heat of Kevlar 49 [16].

Temperature (°C)	Temperature (°F)	Specific Heat, C_P (J/kg °C)	Specific Heat, C_P (Btu/lb °F)
0	32	1220	0.292
50	122	1600	0.383
100	212	1990	0.476
150	302	2360	0.565
200	392	2620	0.626
250	482	2740	0.654
300	572	2840	0.679

Table 3-16. Smoke Generation and Vertical Flammability of Fabrics of Kevlar 49.

Fabric Style Number	Thickness (mils)	Thickness (mm)	Smoke* Maximum Specific Optical Density	Burn Time (sec)	Vertical Flammability† Drips	Glow Time (sec)	Burn Length (in.)	Burn Length (cm)	After Burn Time (sec)
120	4.5	0.11	0	12	None	3.0	1.55	3.94	None
181	11	0.28	6	12	None	6.1	.92	2.34	None
281	10	0.25	7	12	None	5.3	.97	2.46	None
328	13	0.33	4	12	None	6.5	.96	2.44	None

*National Bureau of Standards Smoke Chamber; Flaming Mode.
†Federal Aviation Administration, FAA Part 24, paragraphs 25.853 (A) and 25.853 (B).

Table 3-17. Chemical Stability of Kevlar 49 Yarn and Roving. (Exposure: 24 hours at room temperature, except where noted otherwise.)

Environment	Fiber Strength Decrease	Environment	Fiber Strength Decrease
Concentrated Acids		**Solvents**	
Acetic (99.7%)	None	Carbon tetrachloride	None
Benzoic (3%, 100°C, 100 hr)	26%	Dimethylformamide (DMF)	None
Formic (90%, 100 hr)	7%	Methylene chloride	None
Hydrochloric (37%)	None	Methyl ethyl ketone (MEK)	None
Hydrofluoric		Trichloroethylene	1.5%
(5%)	None	Trichloroethane	None
(48%)	10%	Toluene	None
Hydrobromic (10%, 1000 hr)	60%	**Alcohols**	
Nitric			
(1%, 100 hr)	5%	Benzyl alcohol	None
(70%, 24 hr)	60%	Ethyl alcohol	None
Phosphoric (10%, 100 hr)	1%	Methyl alcohol	<1%
Salicylic (3%, 100°C, 1000 hr)	None	**Other Chemicals**	
Sulfuric		Formalin	1.5%
(1%, 1000 hr)	5%	"Freon" 11 (21 days, 60°C)	2.7%
(10%, 1000 hr)	31%	"Freon" 22 (21 days, 60°C)	3.6%
(70%, 1000 hr)	59%	Gasoline	None
(96%, 24 hr)	100%	Jet fuel	4.5%
Concentrated Bases		Kerosene (21 days, 60°C)	None
Ammonium hydroxide	None	Oil, lubricating	None
Potassium hydroxide	25%	Oil, transformer (21 days, 60°C)	None
Sodium hydroxide	10%	Water, salt (NaCl solution)	<0.5%
Solvents		Water, sea (New Jersey; 12 months)	1.5%
Acetone	None	Water, boiling (100 hr)	2%
Benzene	None	Water, tap	None

PROPERTIES AND PERFORMANCE REQUIREMENTS 91

Table 3-18. Thermal Stability of Kevlar 49 [16]

Property	Value
Long-term use temperature in air	160°C
	320°F
Decomposition temperature	500°C
	930°F
Tensile strength	
At room temperature (16 months)	No strength loss
At 50°C in air (2 months)	No strength loss
At 100°C in air	460,000 lb/in.^{2}
	3,170 MPa
At 200°C in air	395,000 lb/in.^{2}
	2,720 MPa
Tensile modulus	
At room temperature (16 months)	No modulus loss
At 50°C in air (2 months)	No modulus loss
At 100°C in air	16.5×10^6 lb/in.^{2}
	113,800 MPa
At 200°C in air	16.0×10^6 lb/in.^{2}
	110,300 MPa

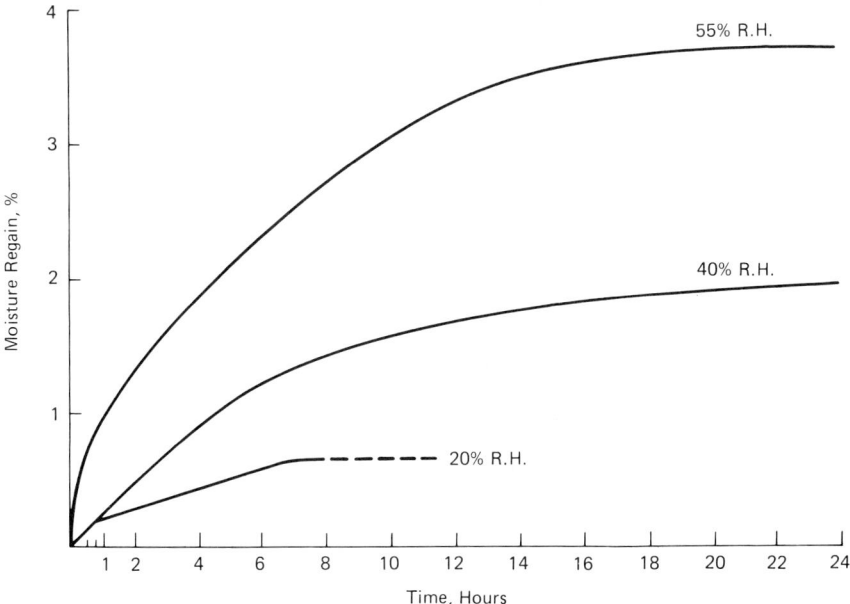

Figure 3-2. Moisture regain at different humidities: Kevlar 49, 380 denier yarn [18].

Table 3-19. Ultraviolet Stability or Kevlar 49 Aramid [16].

Kevlar 49 Product Form	Break Load		Strength Retained (%)
	(lb/in.)	(kN/m)	
Fabric woven from 380 denier (420 DTEX) yarn, 4.5 mil (0.11 mm) thick			
Unexposed	300	52	—
Florida* sun, 5 weeks	154	27	51
	(lb)	(N)	
Three-strand rope, $\frac{1}{2}$ in. (13 mm) diameter			
Unexposed	14,400	64,050	—
Florida sun, 6 months	13,000	57,830	90
12 months	11,600	51,600	81
18 months	9,950	44,260	69
24 months	9,940	44,220	69

Table 3-20. 0° Mechanical Properties, Kevlar 49/Epoxy [16].

Property	0°
Tensile strength, ksi/MPa	200/1379
Tensile modulus, Msi/GPa	11/76
Tensile strain (elongation), %	1.8
Compressive strength, ksi/MPa	40/276
Compressive modulus, Msi/GPa	11/76
Shear (interlaminar), ksi/MPa	6–12/41–83[†]
Shear (in-plane), ksi/MPa	8.7/60
Shear modulus, Msi/GPa	0.3/2.07
Poisson's ratio	0.34
Impact strength, Charpy, (ft-lb/in.^{2})/(J/cm^2)	150/31.5
Creep at 80% ult (1000 hr)	0.05%
Stress rupture (median life)	
90% Ultimate tensile strength	0.80 hr
87% Ultimate tensile strength	2.80 hr
84% Ultimate tensile strength	12.5 hr
80% Ultimate tensile strength	150.7 hr
Fatigue strength (10^6 cycles), ksi/GPa	130/895
Vibration damping ratio	0.86×10^{-2}

*All data are for typical 60% fiber volume laminates with density of 0.05 lb/in.^{3} (1.38 g/cm^3).
[†]Highly matrix dependent; see reference 2 or 16, or check with material suppliers for specific data.

PROPERTIES AND PERFORMANCE REQUIREMENTS 93

Table 3-21. Off-axis Properties of Kevlar 49/Epoxy [16].

Property	90°	0°/90°	±45°
Tensile strength, ksi/MPa	4.3/29.6	91.7/632	14.1/97
Tensile modulus, Msi/GPa	0.8/5.52	5.6/38.6	0.95/6.6
Tensile elongation, %	0.6		
Compressive strength	20/138	29.4/203	18.3/126
Compressive modulus	0.8/5.52	5.6/38.6	1/6.89
In-plane shear		15/103	28/193
In-plane modulus		0.34/2.34	3.04/21

*fv = 60%, 0/90 and ±45 are 50% in either direction.

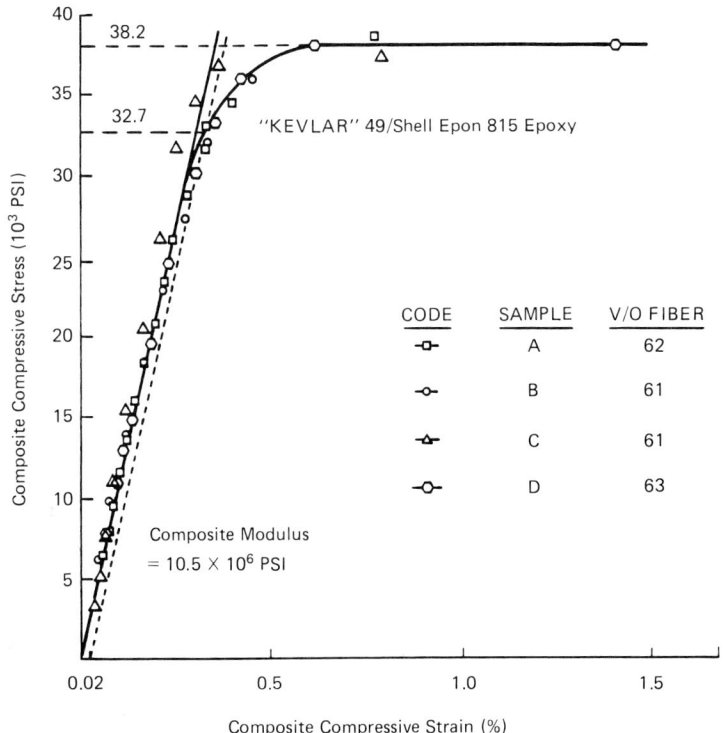

Figure 3-3. Compressive stress-strain Kevlar/epoxy [16].

Table 3-22. Electrical Properties of Kevlar Composites [16] (all composites are unidirectional).

Property	Value	
(Matrix)	At Room Temperature	At 400°
Dielectric constant		
Epoxy*	3.29	3.77
Polyester[†]	3.43	3.60 (300°F)
Polyimide[‡]	3.50	3.64
Loss tangent		
Epoxy*	0.023	0.059
Polyester[†]	0.015	0.028 (300°F)
Polyimide[‡]	0.006	0.013

*Shell "Epon 828."
[†]Bakelite "17449."
[‡]Du Pont "NR-150A."

linear stress-strain relationship in the flexural mode. The onset of nonlinear response occurs above 40,000 psi.

Similarly, Lagasse [18], du Pont [16], and Lubin [2] provide very good reviews of the fatigue response, short fiber reinforced composites, and environmental life of Kevlar composites. The reader is referred to these texts for more in-depth information on the mechanical properties of Kevlar 49 composites.

Table 3-22 shows typical electrical properties for Kevlar 49 composites. Kevlar is competing quite successfully with glass reinforcements for radar transmission structures, particularly where the design critical factors are specific strength or stiffness.

The low to zero coefficient of thermal expansion and the low conductivity of Kevlar 49 composites are attractive for certain applications. Representative thermal properties are presented in Table 3-23 for Kevlar 49/epoxy composites.

CARBON/GRAPHITE REINFORCED COMPOSITES

Patents were issued in the early 1960s in both Japan and England for high modulus, high strength, low density carbon (graphite) fibers. This development was the result of the recognized need for these properties in a reinforcement for building aerospace hardware. In similar fashion to the fiberglass industry, the carbon/graphite materials suppliers have tried to anticipate the industries needs and supply fibers (as well as matrices and composite systems) which would best satisfy many requirements. Therefore, many different products are available, with the result that classification and data presentation are difficult. Even the designation carbon or graphite is still confusing. No attempt at definition will

PROPERTIES AND PERFORMANCE REQUIREMENTS 95

Table 3-23. Thermal Properties of Kevlar Composites [16]

Property	Orientation	Value
Coefficient of thermal expansion		
($\times 10^{-6}$), (in./in. $-$ °F)/(cm/cm $-$ °C)	0°	-4 to -2.3/.02 to $-.01$
	0°, $\pm 45°$ (50:50)	-3 to -2/$-.012$ to $-.008$
	$\pm 45°$	-1.5 to -1/$-.006$ to $-.004$
	$\pm 60°$	34/.14
	90°	35–69/.14–.28
Resin only		65/.26
Thermal conductivity,	0°	.086/.00034
(Btu-ft/hr-ft^2 $-$ °F)/(cal-cm/sec-cm^2 $-$ °C)	90°	.01/.004
(fv = 54.0%)		

be made herein. Instead the terms carbon, graphite, or carbon/graphite (probably most universally correct) will be used interchangeably for all grades of carbonaceous fibers used as reinforcement in advanced composite structures.

Carbon/Graphite Fibers

Table 3-24 is a comparative summary of properties and physical characteristics of carbon/graphite fibers, available from U.S. suppliers. This listing is limited to fibers of major interest for reinforcing organic matrices. It presents information on 30 different fibers. Of these, three are currently considered as the industry standards. They are the high strength (450 to 470 ksi) and lower modulus (33 to 34 Msi) fibers : Union Carbide's "T-300," "BASF" "Celion 3000," and Hercules' "AS." Hysol-Grafil and Hitco have also introduced comparable products: "Grafil X-AS" and " Hi-Tex," respectively. These "standards" have the lowest price and represent the largest structural market. They have the lowest price both because they are used in the largest volume and their lower modulus requires less processing energy than the higher modulus materials.

As shown, many different modulus and strength fibers are available. They are also supplied in different bundle (tow) sizes. These different sizes are achieved by varying the number of strands gathered together into the tow. As shown in Table 3-24, 3000, 6000, and 12,000 filaments per tow are becoming standard for unidirectional lamination and filament winding. To provide additional versatility in weaving cloth, a 1000 filament tow is also supplied.

For purposes of discussion herein, the fibers are classified into four modulus level groupings. The "standard" fibers at $33/34 \times 10^6$, the recently introduced "high strain" graphite at $35/45 \times 10^6$, "Hi Modulus" material at $50/55 \times 10^6$, and the "Ultra-Hi Modulus" at $70/75 \times 10^6$. The higher modulus, greater than 75×10^6 psi materials are also new, have not been widely investigated, and thus are not discussed here.

Table 3-24. Comparison of Carbon/Graphite Fibers, U.S. Suppliers [20]*.

Fiber Producer and Product Identification	Tensile Strength (psi × 10³)	Tensile Modulus (psi × 10⁶)	% Elongation at Break	Coefficient of Thermal Exp. (in./in. °F × 10⁻⁶)	Density (lb/in.³)	Standard Tows (Filaments × 10³)	Filament Diameter (m × 10⁻⁶)
		Union Carbide [21 thru 34]					
T-300 (current industry standard)	470	33.5	1.5	−0.3	.064	1,3,6,12	7
T-500	530	35.0	1.5	−0.3	.065	3,6,12	7
T-600	600	35.0	1.7	−0.3	.065	6	7
T-700	660	36.0	1.8	−0.3	.065	6	6
T-800	(data currently unavailable)						
P-55	250	55.0	0.5	−0.5	.072	2,4	10
P-75S	300	75.0	0.4	−0.7	.072	2	10
P-100	325	105.0	0.31	−0.9	.078	2	10
P-120	325	120.0	0.27	−0.9	.079	2	10
		CCF, Inc. of Badische Americas Corp. [35 thru 42]					
Celion 1000	470	34.0	1.4		.064	1	7
Celion 3000 (current industry standard)	515	34.0	1.5		.064	3,6,12	7
Celion ST	630	34.0	1.84		.064		
Celion G50	360	52.0	0.7		.064	12	6.6
Celion GY-70 and (GY-70S.E.)	270	75.0	0.38		.071	0.384	8.4
		Great Lakes Carbon Corp. [43 and 44]					
Fortafil 3	400	30.0		−.06	.0625	40,160	
Fortafil 5	350	48.0		−.28	.065	40,160	

PROPERTIES AND PERFORMANCE REQUIREMENTS

Panex 30	375		.063	160,320
Stackpole Fibers Co. [45]				
	32.0	1.3		
Hercules Inc. [46 thru 52]				
AS-1 (current industry standard)	450	33.0	1.32	.065	10	8
AS-4	520	34.0	1.53	.065	3,6,12	8
AS-6	600	35.3	1.65	.0657	12	
IM-6	620	42.0	1.60	.063		
HMS	320	50.0	0.58	.066	10	8
HMU	400	55.0	0.70	.067	1,3,6,12	8
Hysol Grafil Co. [53 thru 58]				
Grafil XA-S (standard)	450	34.0	1.31	.0646	6,12	7.1
Grafil XA-S "High Performance"	500	34.0	1.45	.0646	6,12	7.1
Grafil XA-S "High Strain"	560	34.0	1.65	.0648	6,12	7.1
Grafil IM-S	450	42.0	1.07	.0635	6,12	7.1
Grafil HM-S/10k	360	50.0	0.73	.067	10	7.9
Grafil HM-S/6k	400	54.0	0.74	.067	6	6.6
...itco Materials Group [59]				
Hi-TEX	450/470	33.0		.065	6,12	
Hi-TEX HS	525/535	34.0		.065	3,6,12	

*Reproduced *in toto* from Advanced Composite Products and Technology, Inc., TR11284. Reference 20 also includes copies of all data sheets referenced.

Table 3-25. Representative Carbon/Graphite Fabrics* [19].

Designation	Weave	Yearn (Tow) Count Warp		Fill	Fabric Wt (oz./yd^2)	Thickness (in. × 10^3)	Fiber
		Bidirectional Fabrics					
F 1T 093	5 Harness satin	17.8	×	17.8	2.74	3.5	1K, T300
FF 3T 168	Plain	8	×	8	3.6	4.5	3K, T300
F 3T 272	4 Harness satin	12	×	12	5.5	6.0	3K, T300
F 3T 282	Plain	12.5	×	12.5	5.7	7.2	3K, T300
F 3C 282	Plain	12.5	×	12.5	6.6	8.0	3K, Celion
F 3T 584	8 Harness satin	24	×	24	10.9	13.5	3K, T300
F 3C 584	8 Harness satin	24	×	24	12.0	14.7	3K, Celion
F 6C 510	5 Harness satin	10	×	10	9.98	12.0	6K, Celion
F 6C 676	Crowfoot satin	15	×	16 dbl.	15.22	18.6	6K, Celion
F 12C 688	Plain	8	×	8	15.7	18.6	12K, Celion
F 5T 699	Basket	9.25	×	9.25	20.0	30.0	15K, Celion
		Unidirectional Fabrics					
F 1T 712	4 Harness satin	46	×	10	4.02	5.2	Warp, 1K, T300 Fill, 225-1/0 Glass
F 3T 496	Plain	35	×	6	9.3	11.6	3K, T300
F 3C 716	Plain	16	×	24	4.70	6.07	Warp, 3K Fill, 150-1/0 Glass
F 3T 782	Plain	40	×	8	10.4	13.0	Warp, 3K, T-300 Fill, 900-$\frac{1}{2}$ Glass

*This listing does not necessarily indicate fabric availability. If there is an interest in any given fabric, Hexcel or any of the other weavers should be contacted. Normally, small quantities of specific fabrics will be supplied for a price and/or if a large market can be identified.

Like glass and Kevlar, graphite fabric is also available in many standardized and specialty forms. Table 3-25 presents some of these. More complete listings may be found in reference 2 or by contacting the commercial aerospace weavers. Again, graphite fabric can probably be obtained in experimental quantities and will be supplied in volume if a sufficient market can be identified.

Carbon/Graphite Laminate Properties

A comparison of room temperature laminate tensile properties is presented in Table 3-26 for the four major catagories of carbon/graphite-epoxy. These data show that the "standard" fibers can provide tensile strengths to 240,000 psi at a 21 million psi modulus in a unidirectional laminate. The values drop to 66.5 × 10^3 psi and 7.5 × 10^6 psi, respectively, for a quasi-isotropic (balanced 0°, 90°, ± 45°) laminate and to 23,000 and 3,000,000 for a ±45° laminate. Perpendicular, 90°, to the fiber orientation, where the resin matrix is controlling,

Table 3-26 Tensile Strength and Modulus of Carbon/Graphite Laminates* — Typical Properties (Modulus Classified).

Property and Fiber Orientation	Standard Fiber (33/34 Msi)		High Strain (35/45 Msi)		Hi Modulus (50/55, Msi)		Ultra-Hi Modulus (70/75, Msi)	
Tensile ultimate, 0°, ksi/MPa	240/1654	[60]	380/2819	[61]	219/1510	[40]	114/786	[2]
Tensile modulus, 0°, Msi/GPa	21.4/147	[60]	22/152	[61]	29.7/205	[40]	45/310	[2]
Ultimate strain, 0°, %	1.1	[60]	1.7	[61]	0.74	[40]	0.2	[2]
Tensile ultimate, 90°, ksi/MPa	8.8/61	[60]	7.9/54	[61]	5.0/34.5	[40]	3.5/24	[2]
Tensile modulus, 90°, Msi/GPa	1.5/10	[60]	1.32	[61]	.97/6.69	[40]	1/7	[2]
Ultimate strain, 90°, %	0.6	[60]	0.62	[61]	0.52	[40]		
Tensile ultimate, ±45°, ksi/MPa	23/156	[60]	36/248	[62]	17.3/119	[59]		
Tensile modulus, ±45°, Msi/GPa	3/21	[60]						
Tensile ultimate, 0°, 90°, ±45°, ksi/MPa	66.5/458	[60]			55.9/385	[40]		
Tensile modulus, 0°, 90°, ±45°, Msi/GPa	7.5/52	[60]			9.4/64.8	[40]		
Ultimate strain, 0°, 90°, ±45°	1.0	[60]			0.59	[40]		

*All properties for 60 to 65% fiber volume, epoxy matrix, room temperature, dry. Most data are from material suppliers' brochures. However, check with material suppliers, or obtain test data covering the specific material and application being considered before finalizing designs.

Table 3-27 Compressive, Shear, Poisson's Ratio, and Density Properties of Carbon/Graphite Laminates.

Property and Fiber Orientation	Typical Properties* (Modulus Classified)			
	Standard Fiber (33/34, Msi)	High Strain (35/45 Msi)	Hi Modulus (50/55 Msi)	Ultra-Hi Modulus (70/75 Msi)
Compressive ultimate, 0°, ksi/MPa	220/1516 [60]	240/1654 [62]	134/924 [40]	96/662 [2]
Compressive modulus, 0°, Msi/GPa	22/152 [60]	21/145 [62]	24.9/172 [40]	45/314 [2]
Compressive ultimate strain, 0°, %	1.3 [60]	1.1 [62]	0.54 [40]	
Interlaminar (short beam) shear, 0°, ksi/MPa				
In-plane shear strength, ksi/MPa	11.1/76.6 [60]	10/68.9 [64]	8.7/60 [40]	5.4 [64]
In-plane shear modulus, Msi/GPa	0.9/6.2 [60]	0.6/4.1 [64]	0.56/3.86 [40]	0.7 [64]
Poisson's ratio, 0°, specimen, in./in.	0.3 [60]	0.3 [64]	0.27 [40]	0.3 [64]
Density (lb/in.^{3})/(g/cm^3)	0.57/1.58 [60]	.058/1.61 [64]	.057/1.59 [40]	.066/1.83 [64]

*All properties for 60 to 65% fiber volume, epoxy matrix, room temperature, dry.

tensile strength and modulus are reported as 8800 psi and 1,500,000 psi. These are representative values; the reader is cautioned to obtain specific test results which relate directly to the fiber and matrix system and to the actual loading, lifetime, and environmental conditions which will be encountered by the part being designed. Reference 2 provides a reasonable discussion of the design data "knock-down" rationale, which takes fatigue, temperature, and environmental factors into account.

At present the "high strain" fibers have not been characterized as extensively as the "standard" fibers. Therefore, there are fewer published data available. The substantially greater properties achievable with the high strain materials will change this situation. It is probable that as production increases, the 380,000 psi tensile strength (22×10^6 psi modulus) possible with this new generation of fibers will result in their becoming the new industry standard.

The higher modulus fibers are being utilized for structures in which a minimum coefficient of thermal expansion is critical, as well as for stiffness critical minimum weight parts. As shown in Table 3-26, composite moduli of 29.7×10^6 and 45×10^6 psi are possible in unidirectional laminates and 9×10^6 to 13×10^6 psi can be attained in quasi-isotropic panels. Other commonly required design properties are presented for the four typical carbon/graphite epoxy systems in Table 3-27. Figure 3-4 illustrates the effect of fiber volume on laminates or filament wound carbon/graphite hardware. In general, a 60 to 65% fiber volume (fv) can be expected in a quality part.

Typical thermal properties of carbon/graphite laminates are given in Table 3-24. The increasing negativity of the coefficient of thermal expansion with increasing modulus of the fibers (and laminates) is one of the major driving forces for the use of ultra-high modulus fibers. This negative coefficient of the fibers makes it possible to design structures in which the coefficient of thermal expansion ϵ approaches zero. Ashton et al. [63] present equations for calculating ϵ from basic laminate properties. Figure 3-5 illustrates the variation of ϵ in the 0° direction with helical winding angle on a filament wound (or laminated) structure. This graph is for Celanese's GY-70 in Fiberite's X-30 resin matrix. Similar data can be generated for all other combinations. These calculations have been proven to be reasonably accurate by experimental determination [63].

Carbon/graphite laminates also offer superior fatigue properties. Lubin [2] presents extensive data showing that these laminates far exceed aluminum or steel in fatigue life. The dependence of fatigue life on temperature and fiber orientation is also examined in some detail in reference 2. In general, fatigue life decreases with off-axis loading and with increasing exposure temperature.

Other properties of carbon/graphite laminates are of engineering interest. Vibrational damping has been shown to be superior to aluminum or steel [65]. Under proper conditions, wear and frictional characteristics, as shown in Tables 3-29 and 3-30, are comparable to lubricated steel on steel. Once more, the

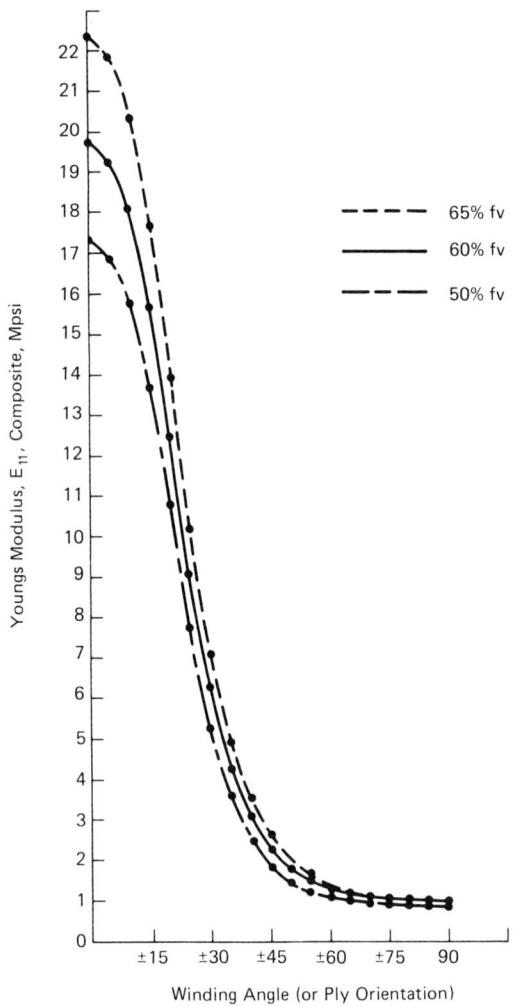

Figure 3-4. Young's modulus graphite/epoxy laminates.*

---

*In-house, first approximation data for "Standard Graphite," Advanced Composite Products and Technology, Inc.

Table 3-28. Thermal Properties of Carbon/Graphite Laminates—Typical Properties (Modulus Classified).

Property and Fiber Orientation	Standard Fiber (33/34 Msi)	Hi Modulus (50/55 Msi)	Ultra-Hi Modulus (70/75 Msi)
Coefficient of thermal expansion, ϵ			
0°, (-10^{-6} in./in./°F)/(10^{-c} cm/cm/°C)	$-0.2/-.36$ [65]	$-0.3/-.54$ [65]	$-0.49/-.88$ [66]
90°, (-10^{-6} in./in./°F)/(10^{-c} cm/cm/°C)	16/29 [67]	14/25 [10]	20.8/37 [66]
0°, 90°, ±45°, (-10^{-6} in./in./°F)/(10^{-c} cm/cm/°C)	1.01/1.82 [65]	0.45/0.81 [65]	
Thermal conductivity			
0°, Btu/hr/ft²/°F/ft	4-5	30-32	[65]
90°, Btu/hr/ft²/°F/ft	0.12	.4-.5	[10]
0°, 90°, ±45°, Btu/hr/ft²/°F/ft	3-4	16-18	[65]
Specific heat, Btu/lb-°F	0.21		

104 ADVANCED THERMOSET COMPOSITES

Table 3-29. Coefficient of Friction, Carbon/Graphite Laminates [69].

Materials (Nonlubricated)	Coefficient of Friction (μ)*
Steel on mid-modulus graphite	0.15 to 0.18
Steel on ultra-high-modulus graphite	0.13 to 0.16
Steel on steel	0.50
Steel on nylon 6/6	0.25
Steel on 20% glass reinforced nylon 6/6	0.18
Steel on 20% graphite reinforced nylon 6/6	0.10
Steel on steel (lubricated)	0.20

*See footnote to Table 3-30

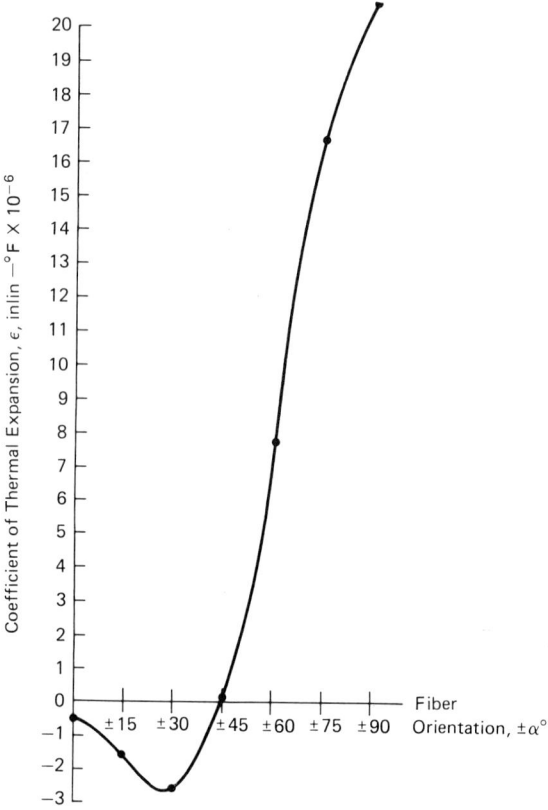

Figure 3-5. Coefficient of thermal expansion, ultra-hi modulus graphite/epoxy laminates [66].

Table 3-30. Wear Resistance of Graphite Composite [69].

Materials (Nonlubricated)	$K\ (10^{-7})$*
Graphite composite against steel	10
Steel against graphite composite	1
Metal against metal (lubricated)	1–10
Steel on steel	5,000
Low carbon steel on copper	5,000
Copper on low carbon steel	10,000
Copper on copper	300,000

*K = year coefficient = $\dfrac{V(\gamma)}{L(P)}$; see references 70 and 71 for discussions of K and μ.

reader is cautioned to run specific tests to verify how the materials to be used perform under the conditions which will be encountered.

OTHER REINFORCEMENT FIBERS

The vast majority of advanced composites are reinforced by glass, Kevlar, and carbon/graphite. Other fibers are available for special purpose applications, and still more are being introduced as a result of improved technology and in response to market needs.

Boron

Boron fibers were introduced at about the same time as graphite. They provide excellent compressive and high modulus properties. However, the relatively high cost of boron reinforcement has restricted it from high volume usage. Ta-

Table 3-31. Boron, Silicon Carbide, and Alumina Fibers.

Property	Boron [72]	Silicon Carbide [73]	Alumina [74]
Tensile strength, ksi/MPa	510/3515	500/3446	200 to 300/1378 to 2068
Tensile modulus, Msi/GPa	58/400	62/427	55/379
Density (lb/in.^{3})/(g/cm^3)			
.004 in. diameter filament	0.093/2.58		
.0056 in. diameter filament	0.090/2.49	.11/3.05	
.008 in. diameter filament	0.089/2.47		
20 microns (200 filaments/yarn)			0.141/3.9

Table 3-32. Boron, Silicon Carbide, and Alumina Laminate Properties.

Property	Boron [72]	Silicon Carbide [73]	Alumina, 0° [75]	Alumina, 90° [75]	Alumina, 0°, 90° [75]
Tensile strength, ksi/MPa	220/1516	229/1578	70 to 80/482 to 551	6 to 10/41 to 69	35 to 45/241 to 310
Tensile modulus, Msi/GPa	30/207	33/227	30 to 33/207 to 227	2.5–3/17 to 21	15 to 16/103 to 110
Compressive strength, ksi/MPa	470/3239	326/2247	330–350/2274 to 2412		170 to 180/1172 to 1241
Horizontal shear strength, ksi/MPa	15.0/103	15/103			
Density		.084/2.33	.097–.101/2.7 to 2.8		
Dielectric constant (RT, 9.26 Hz)			5.5–6.0		

*All data are with epoxy matrices and are for room temperature dry conditions.

Table 3-33. Kevlar/Alumina Hybrid Laminates* [76].

Property	0°	90°	0/90°
Tensile strength, ksi/MPa	130–150/896–1034	2–4/14–28	70–80/482–551
Tensile modulus, Msi/GPa	20/138	0.8/5.5	12/83
Compressive strength, ksi/MPa	125–150/862–1034		80–100/551–689
Compressive modulus, Msi/GPa	18/24		13/90
Short beam shear, ksi/MPa	8–10/55–69		
Shear modulus, Msi/GPa	0.5–0.6/3.5–4.1		
Physical Properties			
Density, (lb/in.^{3})/(g/cm^3)		0.07/1.98	
Dielectric constant (9.26 Hz, RT)		4.5	
Thermal conductivity, cal/cm sec °C (normal to plies)		11×10^{-4}	

*fv = 28 to 30 alumina plug 32 to 30 Kevlar 49.

ble 3-31 summarizes the properties of boron fiber. Table 3-52 shows typical laminate properties.

Silicon Carbide

Silicon carbide fibers also offer superior compressive strengths in epoxy matrices. The fibers offer excellent wetability for metal matrices and are useful to 1,200°C (2192°F). Tables 3-31 and 3-32 also present properties of silicon carbide and silicon carbide/epoxy laminates.

Alumina

Short length alumina fibers (whiskers) have been available for some time. Du Pont has recently introduced continuous fibers with a melting point of 2045°C (3713°F). Alumina fibers are useful for reinforcing metal and organic matrices. Alumina provides high modulus and high compressive strengths in epoxy matrices. Represenative alumina fiber and laminate properties are given in Tables 3-31 and 3-32.

Du Pont has also published data (Table 3-33) which indicate that hybrids of Kevlar and FP fiber, du Pont's trade name for alumina fiber, have good translation of tensile and compressive properties. A 28 to 30 fiber volume (fv) FP with 32 to 30 fv Kevlar provides 125 to 150 ksi in both tensile and compressive strengths.

References

1. Delmonte, J., *Technology of Carbon and Graphite Fibers*, Van Nostrand Reinhold, New York, 1981.
2. Lubin, G., Ed. *Handbook of Composites*, Van Nostrand Reinhold, New York, 1982.
3. L. Lackman, et al., *Advanced Composite Design Guide*, Third ed., AFMI, Contract to Rockwell F. 33615.71.C.1362, 1973. (Note: the reader should contact AFML/WPAFB, Ohio, to be sure the latest volume of this ongoing advanced composites reference series is being considered.)
4. Gross, S., Ed., *Modern Plastics*, McGraw-Hill, New York, Vol. 61, No. 7, p. 54, (1984).
5. Anon., "Textile Fibers for Industry," Pub. No. 5-TOD-8265, Owens Corning Fiberglass Corp, Toledo, Ohio, February 1983.
6. Anon., "Reinforced Plastics by Design," Bulletin F352-R-8112-5M, PPG Industries Inc., Pittsburgh, Pa.
7. Anon., "S2 Glass^R Fiber Bridges the Reinforcement Gap," Pub. No 5-ASP-10139, Owens Corning Fiberglass Corp., Toledo, Ohio, August 1980.
8. Anon., "Astroquartz, The Flexible Approach," J. P. Stevens Co., New York, 1983.
9. Anon., "S-2 Glass^R Fiber, High Performance/Low Cost Reinforcements," Pub. No. 5-PL-10182, Owens Corning Fiberglass Corp., Toledo, Ohio, February 1981.
10. Friend, C. A., J. G. Poesch, and J. C. Leslie, "Graphite Fiber Composites Fill Engineering Needs," Hercules, Inc., Magna, Utah, 1971.
11. Agranoff, J., Ed., *Modern Plastics Encyclopedia*, McGraw-Hill, New York, revised annually.
12. Anon., "Bulletin #98, Fabric Specifications and Competitive Equivalents," Hexcel, Inc., Dublin, Calif., 1972.
13. Anon., Product Data—Industrial Glass Fabrics, J. P. Stevens & Co., New York, #301-3, 1975.
14. Anon., Product Data—Boat and Tooling Glass Fabrics, J. P. Stevens & Co., New York, #302-1, 1975.
15. Anon., "An Introduction to Fiberglass-Reinforced Plastics/Composites," Pub. No. 1-PL-6305-A, Owens Corning Fiberglass Corp., Toledo, Ohio, December, 1976.
16. Anon., "Kevlar^R 49 DATA MANUAL," E. I. du Pont de Nemours & Co., Wilmington, Del.
17. Anon., "Characteristics and Uses of Kevlar^R 29 ARAMID," Bulletin #375, E. I. du Pont de Nemours & Co., Wilmington, Del., 1976.
18. Langston, P. R. et al., Proceedings of TECHNICAL SEMINAR III, Design and Use of KEVLAR Aramid in Aircraft," E. I. du Pont de Nemours & Co., Wilmington, Del., 1981.
19. Hexcel Inc., Dublin, Calif.
20. Leslie, J. C., TR#12184, High Strength-High Strain Graphite Fibers and Prepregs, Jim Leslie C.&S.R., Huntington Beach, Calif., December 1984.
21. through 34. Anon., "Technical Information—Thornel^{TM} Carbon Fiber Data Sheets, T300, 1K through P-120 2K," Union Carbide Corporation, Danbury, Conn., 1983.

35. through 42. Anon., "Material Properties Celion[R] Carbon Fibers Celion 1000 thru Celion GY-70 SE," Celanese Corporation, Chatham, N.J., 1983.
43. through 44. Anon., "Fortafil 3 and Fortafil 5" Data Sheets, Great Lakes Carbon Corporation, New York, 1982.
45. Anon., "Panex 30 Data Sheet, Bulletin 3M-381," STACKPOLE FIBERS Co., Lowell, Mass.
46. through 52. Anon., "Magnamite Graphite Fiber," Data Sheets AS-1 thru HMU (838-1, 847-3, 860, 27100, 840-1, 851), Hercules, Inc., Wilmington, Del.
53. through 58. Anon., "Carbon Fibers, Resins, and Specialty Prepregs, Data Sheets," Hysol Grafil Co., Pittsburg, Pa., 1983.
59. Anon., "Hitco Hi-Tex Structural Carbon Fiber, Product Data Bulletin," Hitco Materials Div., Gardena, Calif., 1982.
60. Anon., "Celion Carbon Fibers, Material Properties Composites, Bulletin DFCCIA," Celanese Corp., Chatham, N.J., 1982.
61. Anon., Preliminary Data Sheet HX1504 (F584); HEXCEL Corp., Dublin, Calif., 1983.
62. Anon., "Data Bulletin 5245," Narmco Materials, Inc., Anaheim, Calif., 1983.
63. Ashton, J. E. et al., *Primer on Composite Materials*, Technomic Publishing Co., Stanford, Conn., 1969, pp. 88–91.
64. Anon., *Fiberite Composite Materials*, Fiberite, Inc., Winona, Minn., 1984.
65. Anon., "Magnamite[R] Graphite Fibers," Hercules, Inc., Wilmington, Del., 1981.
66. Cohen, L., personal communication with J. C. Leslie, ACPT, Inc., March 4, 1982. (Dr. Cohen is with the Smithsonian Institution National Observatory, Cambridge, Mass.)
67. Kuebeler, G. C. and C. E. Jordan, "The New Diet Material for Structural Applications," *Hercules Chemist*, No. 62, Herecules, Inc., Wilmington, Del. (July 5, 1971).
68. Fourney, W. L. and J. G. Poesch, "Dynamic Modulus and Damping in Graphite Composites," Hercules, Inc., Magna, Utah.
69. Anon., "Engineering Applications of Graphite Fiber Composites," Hercules, Inc., Magna, Utah, 1975.
70. Berg, C. A., S., Barta and J. Tirosh "Friction and Wear of Graphite Fiber Composites," *Journal of Research of the N.B.S.*, **76C** (January–June 1982).
71. Brown, R. D. and W. R. Blackstone, "Evaluation of Graphite Fiber Reinforced Plastic Composites for Use in Unlubricated Sliding Bearing," Southwest Research Inst., San Antonio, Tex., 1973.
72. Boron Composite Materials (0481-2-2M), AVCO Specialty Materials Division, Lowell, Mass.
73. Silicon Carbide Composite Materials (0481-20-2M), AVCO Specialty Materials Division, Lowell, Mass.
74. Technical Data: FP Fiber-General (E-56612), E. I. du Pont de Nemours & Co., Wilmington, Del.
75. Technical Data: FP/Epoxy Composites (E-56614), E. I. du Pont de Nemours & Co., Wilmington, Del.
76. Technical Data: FP/KEVLAR 49/Epoxy Composites (E-56615), E. I. du Pont de Nemours & Co., Wilmington, Del.

4
ELECTRICAL AND ELECTRONIC APPLICATIONS

D. P. Seraphim
D. E. Barr
W. T. Chen
G. P. Schmitt

IBM Corporation
Endicott, New York

The most recent, innovative electronic products introduced to the world markets, from the personal computer to powerful multimillion instruction per second computer systems, utilize printed-circuit composites.

Two of the materials commonly used in making these composites, epoxy resin and hardener, are found on almost everyone's shelves. They are used to bond glass cloth and copper into composites, which are laminated, photoformed, drilled, and punched; chemically treated, including both electroless and electrolytic plating; encapsulated; and soldered. The thousands of configurations made annually in the United States could encircle the earth several times with a 1 foot wide band. A recent study [1] from the Institute for Interconnecting and Packaging Electronic Circuits (IPC) estimates that 41 million pounds of resin and 42 million pounds of glass and cloth, a total of 114 million linear yards of laminate, were used in 1982.

Printed-circuit-board volume consumption is driven by, and closely parallels, semiconductor consumption. According to a recent study [2], the 1983 semiconductor worldwide consumption was $23.9 billion, with computer and office equipment the largest and fastest growing segment, followed by consumer electronics and communication. This market is expected to increase 36% in 1984 and 22% in 1985. Other studies [3, 4] have shown that the corresponding printed-circuit-board consumption was worth $3.6 billion in 1983, with a cumulative industry annual growth of 17% leading to a 1988 projection of $7.8 billion in the United States alone.

Based upon dollar values, about 92% of U.S. printed-circuit boards are in rigid composites and 8% in flexible circuits. Double-sided glass laminate is projected to dominate the market until 1987, with a current share of 52%. Multilayer composites, however, will show the greatest growth, accounting for 29% of the U.S. printed-circuit-board production. Coupled with the growth in multilayer will be the introduction of new technological innovations in design, materials, performance, and production processes and equipment.

Semiconductor and circuit packaging constitutes the heart and backbone of the electronic equipment industry, which has been experiencing a 14 to 15% annual sales increase. Moreover, this technology is becoming increasingly important in a number of nontraditional electronic industries, such as the automotive industry [5].

The electronic products manufactured in 1984 range from a simple one-layer configuration, through configurations with fewer than 200 circuit interconnections on two layers, to highly complex 20 layer arrangements [6, 7] that contain kilometers of copper interconnection the width of a human hair. An outstanding example of one of the more simple composites produced in high volume is Ford Motor Co.'s EEC-IV module [5], which contains the program and data for engine control, a memory-retaining instruction when the engine is off, a drive for multiple power transitions (fuel injection), a central processing unit (40 terminals), and a few other components as well as about 50 "discretes" (resistors and capacitors, etc.).

A wide range of basic fabric materials, along with various resins, adhesives, and films, are processed with increasingly sophisticated equipment such as lasers, continuous photoforming equipment [8], and huge, automatic wirebonding equipment. To meet the ever growing demand for electronic printed-circuit-board products, a revolution is happening in the automation of component soldering [9] and assembly for the composites.

The components for large computer systems are assembled into application-driven designs, intensifying the need for miniaturization that combines both performance and function into the smallest possible space [10, 11]. However, at the opposite end of the system spectrum, it is important that all the components of a personal computer—keyboard, display, printer, and base system—reside conveniently on a table top in an arrangement that is compatible with the individual user.

A view under the dash of an automobile illustrates the reason for using the smallest possible computer components. In addition to conserving space, miniaturization results in added performance. The trend in the highest performance computer systems for large business and scientific usage is to integrate the maximum function on semiconductor chips. This enables the signals to travel the shortest possible distance, only thousandths of an inch, at the speed of light, 3×10^{10} cm/sec modified by the dielectric constant. Nonetheless, whenever a small fraction of the signals leave the chips, traversing the organic-copper laminate to other chips, the order of magnitude caused by longer distances results in a significant performance delay. Therefore, the performance-oriented system designer needs innovative "packaging" of component technologies [11, 12] that will support many chips placed close together.

The electronic packaging designs must provide the interconnections (wiring) while also providing the proper electrical, thermal, and mechanical environment for the components. Since increasing either the semiconductor chip function or the spacing density will increase the power density, there are many innovative approaches [11, 13] for providing voltage distribution and heat extraction. Micromechanical techniques for analyzing the designs and interfacial configurations [14] are becoming, therefore, increasingly sophisticated. The state of the art in plastic molding of intricate connector housings is being driven by the need for high density connectors [15, 16], which are an integral part of the electronic package design.

This chapter explores the basic elements of electronic package designs, including packaging hierarchy, focusing on the broad range of applications presented in the introduction. In addition, the basic materials and processing for both simple and complex composites will be explored, followed by a brief discussion of the multidisciplinary sciences that contribute to the process of developing printed-circuit composites.

ELECTRONIC PACKAGING HIERARCHY

The major hierarchical elements of electronic packages are semiconductor chips, modules (chip carriers), printed-circuit cards or planars, printed-circuit boards, and cables. These are often assembled in a characteristic arrangement (Figure 4-1) when all are used together. In this section, each component will be described in turn, along with the rationale for the various levels of complexity in the packages used in the industry. Although there is a continuum of complexity, we will try to separate the low-end applications from the high-end.

Low-end applications use millions of panels per year to make microprocessors* which provide a myriad of services. High-end applications, at the opposite end of the spectrum, are shipped in lower volume but require greater reliability and performance.

---

*A microprocessor is actually a miniature computer that has all the elements of architecture to handle programs, storage instruction sets, etc. See *Micro-processors/Microcomponents An Introduction* by Donald D. Givone and Robert D. Pioesser, McGraw-Hill, 1980.

ELECTRICAL AND ELECTRONIC APPLICATIONS

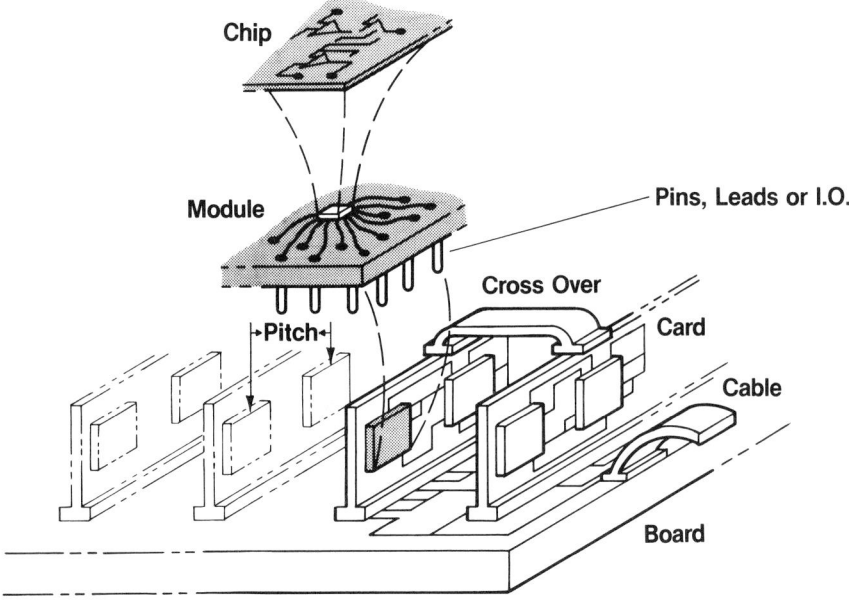

Figure 4-1. Frequently used packaging components in one packaging hierarchy. The cards with appropriate contacts are usually plugged into connectors in the board.

Typical low-end eight-bit microprocessor chips contain 12 to 40 signal input and output terminals, IO. These chips are available in components (chip carriers or modules). They are (see Figure 4-2): flat packs, pin grid arrays (PGA), leadless chip carriers (LCC), plastic leaded chip carriers (PLCC), small outline integrated circuits (SOIC)®* and dual in-line packages (DIP). Flat packs, pin grid arrays, and dual in-line packages have been available for approximately 20 years, while the other components are recent innovations. For a comprehensive discussion of the various components available and packaging technological trade-offs, see *VLSI and the Substrate Connection, the Technological Tradeoffs of the Package Board Interface* by J. W. Balde and D. Brown [17].

Two Signal-layer Composites

A card (Figure 4-3) or small planar with two interconnection layers is an excellent example of the most commonly used composite. With just two tracks per 0.100 inch grid (equalling 20 tracks per inch), such a card can interconnect

---

*Registered trademark of Motorola, Inc.

114 ADVANCED THERMOSET COMPOSITES

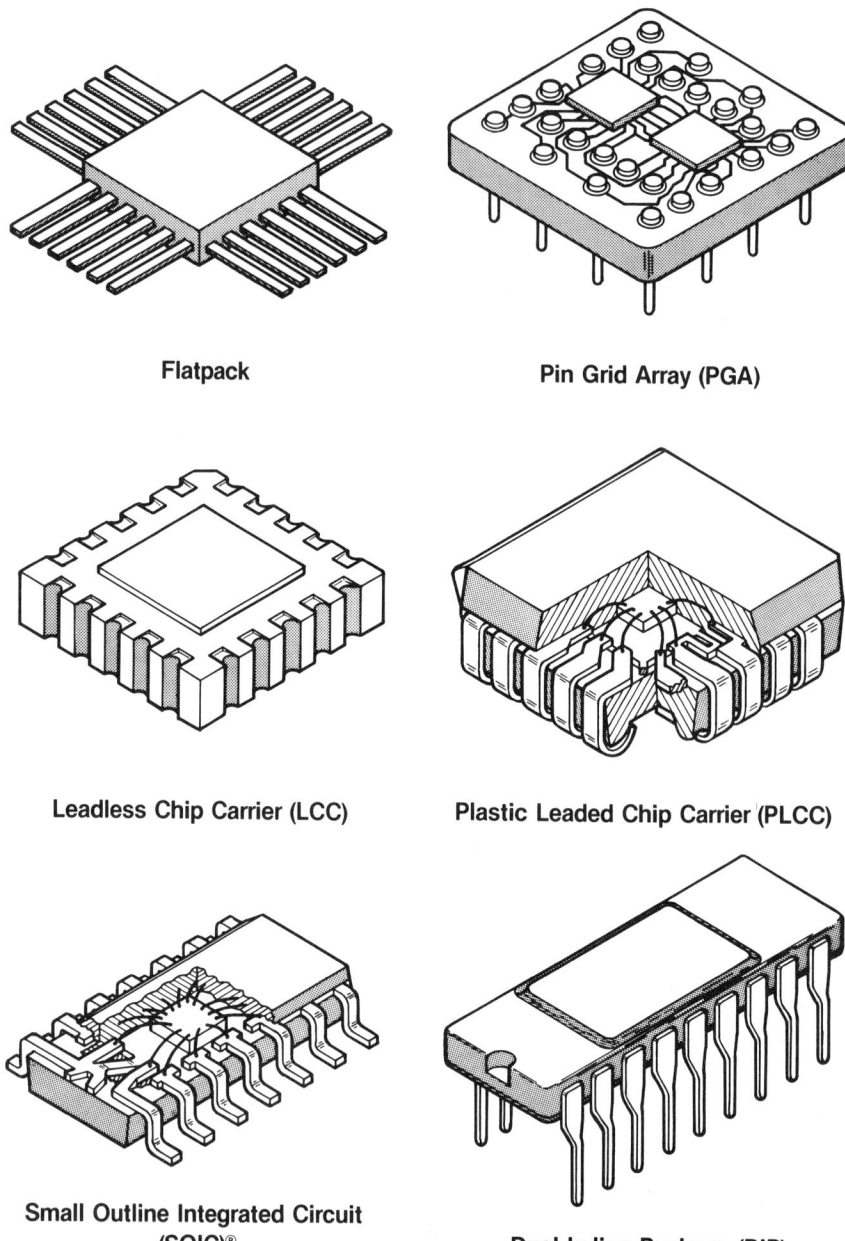

Figure 4-2. Chip carriers or modules (first-level packages) — common components which hold semiconductor chips and supply the connections or leads between the chip and printed-circuit composites.

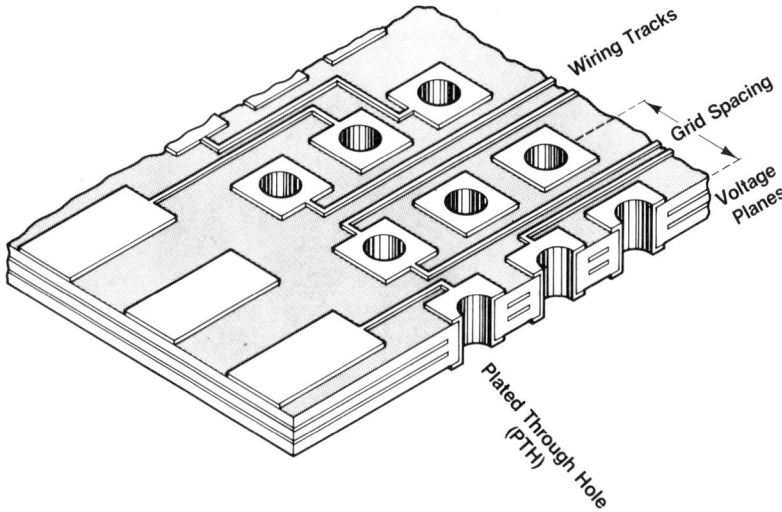

Figure 4-3. Composite with two signal-interconnection layers, plated through holes, and voltage reference planes. There are two wiring tracks between grid sites (two lines per channel or 20 tracks per inch).

about ten components (i.e., a microprocessor). These components could be evenly spaced in a 7 inch by 3 inch composite in a 5 by 2 array. Using this density, 60 tracks (or lines) are located in the width (3 × 20) with a length of 7 inches per track, a total of 420 (60 × 7) inches per layer of composite. A simple algorithm [18], as shown in Figure 4-4, calls for approximately 2.25 inches of interconnection per lead or IO, multiplied by the component spacing (or pitch P). For this example, the average spacing, or pitch (P), is about 1.4 inches (7 inches divided by 5) per component. In our example, we have a total of 240 IO to interconnect (eight chips with 20 IO and two chips with 40 IO). The required wiring is, therefore, about 750 inches. The two layers, each with a capacity of 420 inches, exceed the requirement of 750 inches, an acceptable solution.

A two-layer interconnection design of this complexity is typical of the eight-bit microprocessor and memory printed-circuit composites that were constructed in 1984. These may be used as single cards or planars without the total hierarchy shown in Figure 4-1. Figure 4-3 illustrates a printed-circuit composite with this type of interconnection configuration and includes two voltage planes. As shown in the figure, a connection is required for layer-to-layer communication. This connection is achieved with a hole, usually mechanically drilled, which has copper-plated walls. The hole also provides a receptacle for the com-

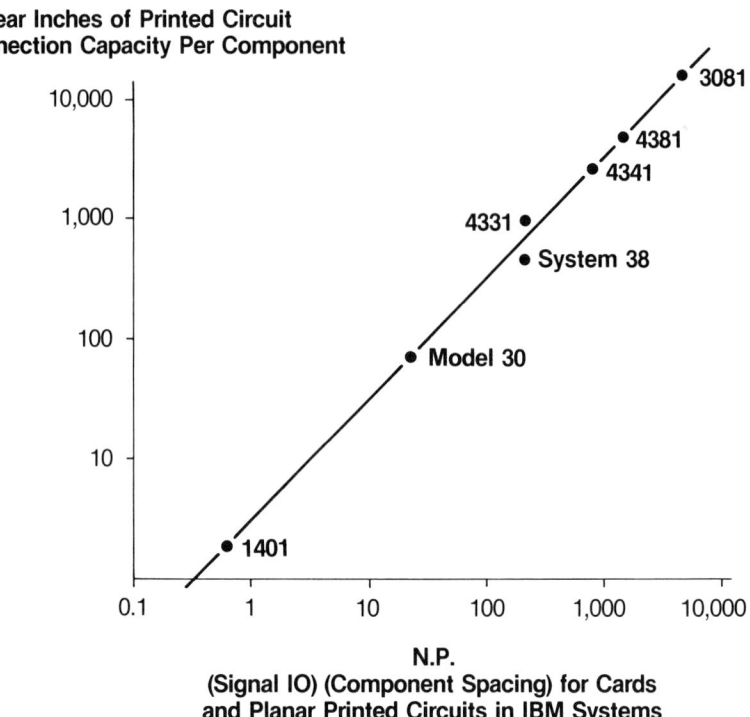

Figure 4-4. Approximate interconnection length required for printed circuits connecting components with N leads with component spacing P. $L = 2.25NP$. The IBM systems shown on the graph include: the IBM 1400 Processor System; IBM 360 Processor System Model 30; IBM 4300 Processor System models; IBM 4381 Processor System, Model 38; and the IBM 3081 Processor System.

ponent leads of DIPs and PGAs, which are soldered into the printed-circuit composites. This two-sided type of design with and without voltage planes and the simpler one-sided designs are, in fact, still the most common printed-circuit configurations used to date (1984) [1]. This design still accounts for approximately 80 to 90% of the world's production of laminate, with sizes smaller than the 7 inch by 3 inch example noted earlier to areas as large as 17 inches by 22 inches.

The larger composites, often called planars, have 150 components or more for the expansion of memory and logic function, all on one panel. These almost always require buried-voltage distribution planes. The larger size composites, which are equivalent to two or more smaller composites, save connectors and cabling between two planars, which result in increased performance and low-

ELECTRICAL AND ELECTRONIC APPLICATIONS 117

Figure 4-5. Two printed-circuit composites used in the IBM Personal Computer.

ered costs. The more advanced personal computers use these large planars, as shown in Figure 4-5, and some personal computers even contain several large planars, with connectors and cables joining them. The components may be spread out and require only two signal layers for wiring, or they may be tightly integrated and require four layers of interconnection.

Multilayer Composites

Architecture for business computers more powerful than personal computers typically calls for logic chips with a much larger number of IO [18, 19]. In fact, the need for IO grows expontentially with the number of circuits (C) to the two-thirds power [18], IO = $\alpha C^{2/3}$ where α is a constant about 2. For example, in 1979, the pin grid array components that were announced for the

118 ADVANCED THERMOSET COMPOSITES

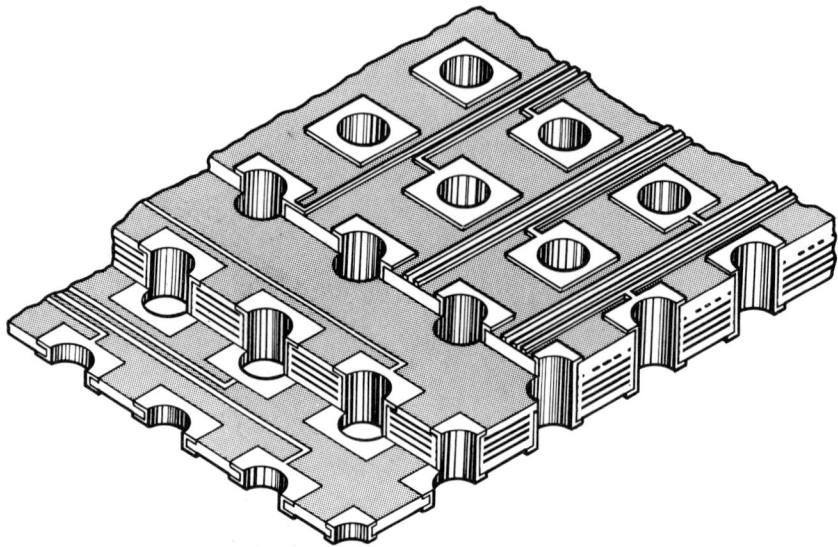

Figure 4-6. Composite with four signal-interconnection layers and three tracks per grid site.

IBM System 38 processor, were 28 mm square and contained more than 100 pins to support an average of 500 circuits per chip. The composite printed-circuit layers required to package these components, and the density of the interconnections are approximately three times more complex—30 tracks per inch and four layers of interconnection instead of two layers—in comparison to the planar composite discussed earlier.* This increased complexity is in direct proportion to the IO or pin count per components, as noted earlier. Furthermore, the increased circuit density requires an additional two planes of copper used for power distribution.

Multilayer printed-circuit composites of this and slightly less complexity are typical of midrange computer systems produced by numerous vertically structured computer manufacturers in the early 1980s.

Cross sections of the composites are shown in Figure 4-7 to illustrate the design features, and a perspective (Figure 4-6) is provided to show the effect on component assembly spacing. In Figure 4-7, the three structures are illustrated with equal interconnection capacity. In structure A, a large area is used, with widely spaced interconnections at 10 tracks per inch. In structure B, the

---

*To compare the composite printed-circuit layers and the planar composite, see Figures 4-3 and 4-6. Also compare IBM Processor System 360, Model 30 in Figure 4-4, which primarily used 16 leaded components, to IBM Processor System 4331, Model 38, which used 100 leaded components.

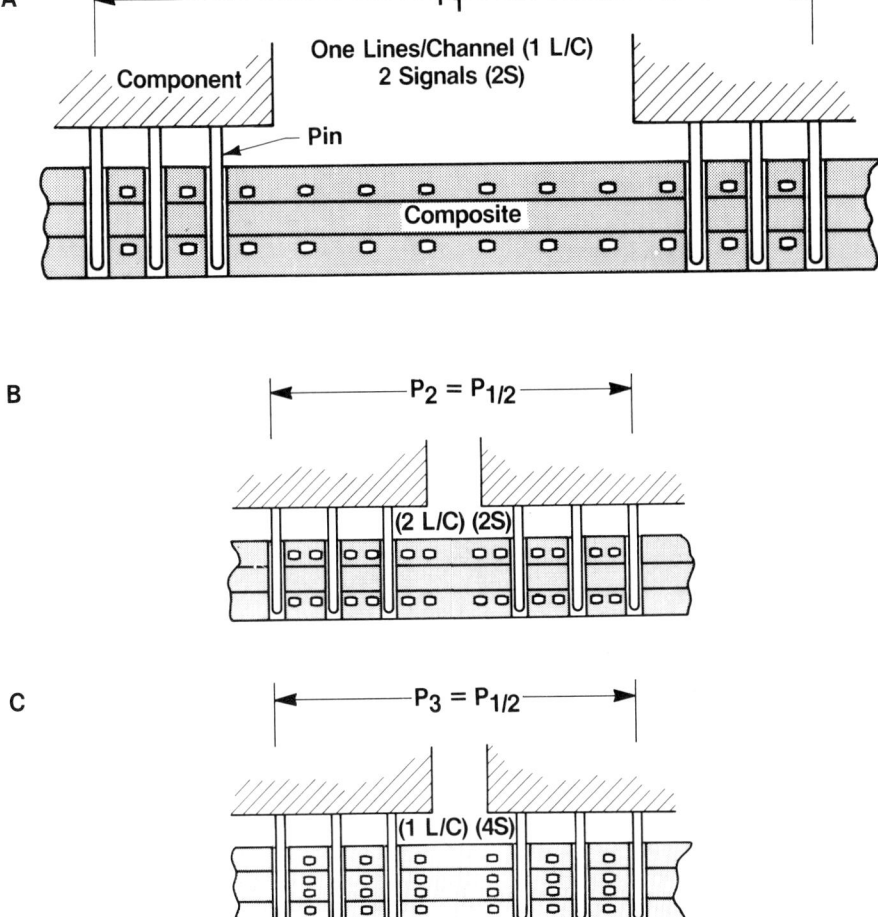

Figure 4-7. Effect of composite interconnection density on component spacing.

spacing between interconnections is reduced to provide 20 tracks per inch. In case C, the layers are doubled over A to provide 20 tracks per inch. In cases B and C, the spacing between the components may be reduced by one-half. Therefore, since the distance traveled by the signals is significantly decreased, the performance is improved accordingly. However, in case B, the total area used for printed-circuit layers is substantially less than in cases A and C. If the process technology is presumed to be adequate and yields are comparable, B would be the superior option for combined price and performance. Case B might also be the best selection for reliability in view of the shorter total interconnection length, the simpler plated through hole structure with fewer layers, and the decreased process steps, in comparison with case C. This is a simple illustration of the keys to performance, cost, and reliability in printed-circuit composites. It demonstrates the movement toward microminiaturization of interconnection and spacing through either photolithographic technology or the addition of multilayers, which require fabrication of tiny, plated through holes (PTH) to effectively interconnect the layers.

Card-on-board Package

In the event that the electronic functions exceeds the amount that can be suitably packaged on one panel, two panels may be joined by connectors and cables. However, if many panels must function together, the planars or cards may be plugged directly through connectors into a back panel composite or board (Figure 4-1), which is a complex multilevel printed-circuit composite in itself. In this structure, the electronic package becomes a three-dimensional package of planars (cards)—many in a row and parallel—joined on their edges to a printed-circuit panel composite (board) by connectors. This type of card-on-board package, as shown, was used in the late 1960s and 1970s to build increasingly larger computers in the smallest possible spaces. Systems containing hundreds of boards have been constructed, with the board-to-board connections cabled together, as shown in Figure 4-8

To recap, a large hierarchy has been presented: chip, component, planar or card, connector, back panel or board, and cable. In recent applications, the cards may contain a very large number of contacts at one end (Figure 4-9), or both ends may contain connectors and join to back panels on both ends [20]. This might be called a shoe-box or a three-dimensional package, with the cards sliding into a zero-insertion-force connector·(ZIF) in the box.

The capacity for even higher densities is obviously limited in this configuration by the rather large spacing between components, which contain single chips, on the card or planar. The spacing between cards is also dependent upon the component size, the connector density (Figure 4-9), and perhaps other factors like cooling, which requires the circulation of a large volume of air over the card with its components. Furthermore, a large surface area at the component level may be necessary for convective heat transfer to the air-cooling system.

Multichip Modules on Planar Package

An alternative approach is to do away with the component package (PGA, DIP, PLCC, SOIC, etc.) and deal with a multichip component package. The point of this approach is to reduce the hierarchy to chip, multichip module, and

ELECTRICAL AND ELECTRONIC APPLICATIONS 121

Figure 4-8. Computer gate showing dense cabling between many boards.

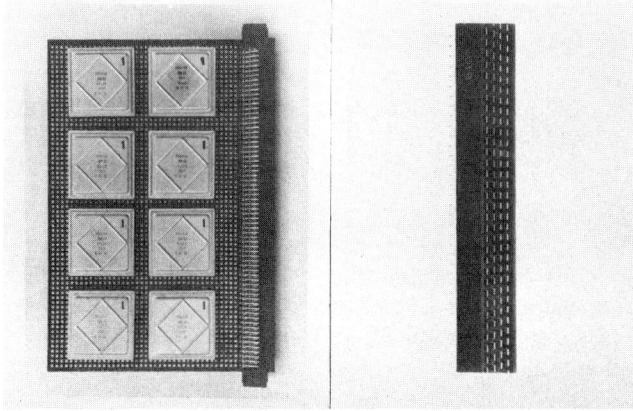

Figure 4-9. Card containing eight PGAs with dense connector blades (268 contacts), which plugs into a mother board with as many as 18 additional cards.

printed-circuit planar. Without the component items (pins, cap, etc.), the chips could be positioned next to each other on the multichip module [11, 12].

In 1980, IBM announced such a system, with four very large printed-circuit boards [6, 7] that individually interconnected six or nine of these modules. Each module was able to interconnect a hundred chips. Since the modules are very large (4 inches × 4 inches), the module spacing, or pitch, was even larger and the interconnection length on the printed-circuit composite, per IO, had to be approximately 8 inches. This size factor, plus the 1800 IO per module, created a need for the largest and most complex (see Figure 4-4) printed-circuit composites in the industry to date (1984). As in most of the industry, these composites were made of epoxy-glass. The interconnection requirements (Figure 4-4) were 40 tracks per inch in six signal layers containing approximately a mile of photoprinted copper interconnection only .0035 inch wide. Twelve photoprinted copper voltage planes were laminated into the structure for power distribution, delivering several hundred amps to the modules. A representative cross section, shown in Figure 4-10, depicts the complexity at high magnification. The figure illustrates the relationship of the circuitry to the laser-drilled plated through holes, .006 inch in diameter, and the 40,000 high aspect ratio plated through holes, .016 inch in diameter and .180 inch long. Each thermally cooled module (TCM) plugged into 1800 connectors [15, 16] that were soldered into the plated through holes in the printed-circuit composite.

If the packages, which allow the chips to be closely spaced, result in the highest performance, it would seem that the TCM is most effective at achieving the shortest interconnection distances. The drawback of a high dielectric constant for the ceramic and a limited interconnection spacing is balanced by the ease of laminating 30 layers into a composite structure. The printed-circuit board, however, is rendered highly complex by the demand for many long interconnections. The development costs in shrinking the 100 printed-circuit boards with tens of thousands of cables (Figure 4-8) that were used in every system, to a simple hierarchy of chips, modules, and four boards, is certainly worthwhile in terms of gains in performance, cost, and reliability. The key to success in this venture is the adaptability of the epoxy-glass system to multilayer printed-circuit fabrication.

Surface-mount Packages

The increasing capabilities available on semiconductor chips result in an ever increasing demand for more IO to effectively utilize this chip function. Microprocessor chips are progressing from 8 to 16 bits and even to 32 bits in the more advanced technology while the circuits per chip are increasing to thousands. The need for IO has increased proportionately with the need for circuits. Thus, where 40 to 60 IO were required for the upper end of the component

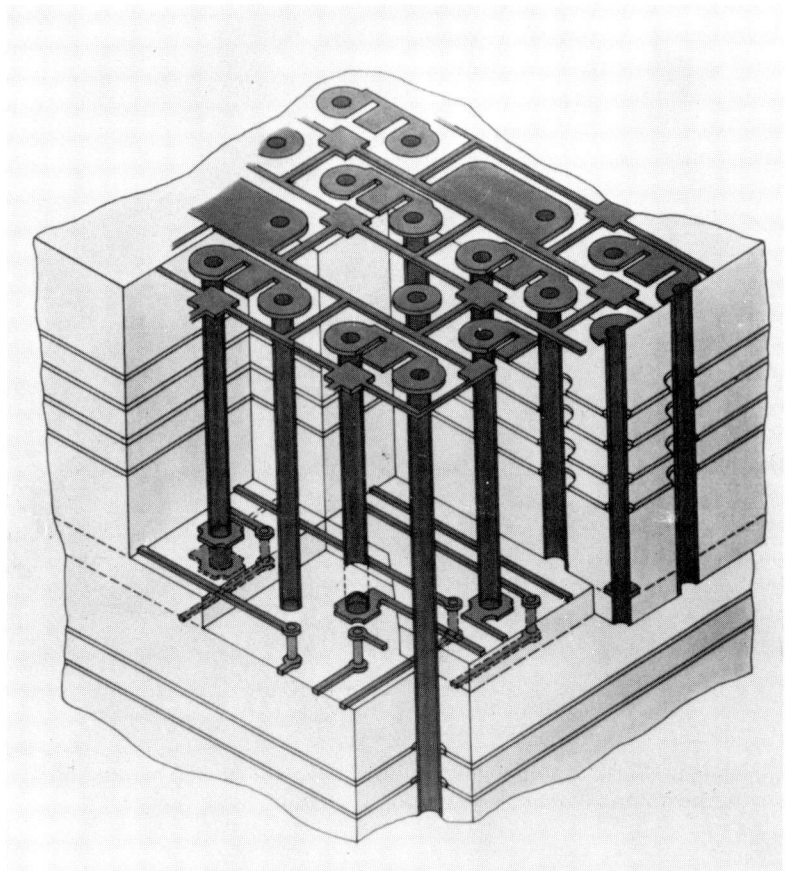

Figure 4-10. A view of several grid sites of a 20-layer printed-circuit composite. The layers are shown to illustrate the location of programmable vias connecting one of the three pairs of signal planes.

function, several hundred IO are now needed. To answer this need, the industry is developing components with terminals on all four sides, PLCC. This is a contrast to DIP, in which the leads are located only on two sides. This four-sided arrangement allows the component to be much smaller, while generally increasing the IO component capability in the industry.

Since the printed-circuit composites must have an entry into the wiring planes for each IO, a very high density of holes is necessary when, for example, 200 IO chips are used. If the chip spacing must be contained in a limited area, 200 holes would be needed in the same space in the composite. In addition, at least one extra hole, sometimes more, is required for wiring efficiency [21] to allow communication between X-direction and Y-direction tracks. If we use the pre-

124 ADVANCED THERMOSET COMPOSITES

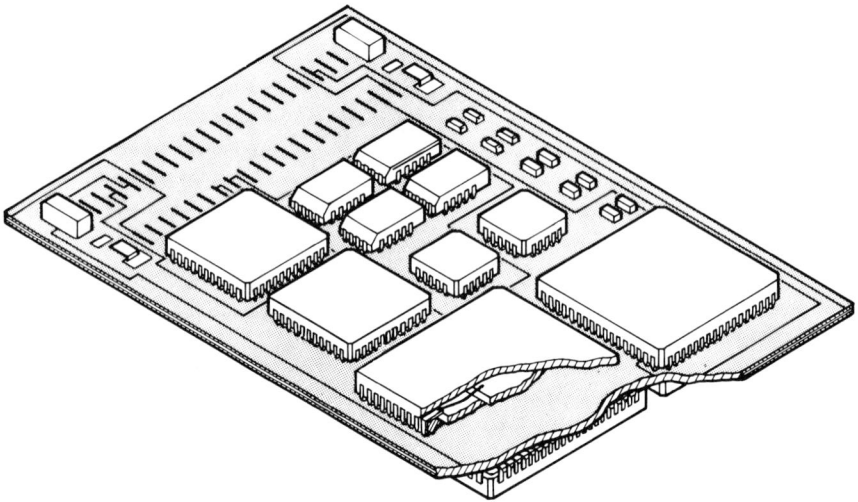

Figure 4-11. Surface solder cards with plastic leaded chip carriers (PLCC). As shown, the components can be mounted on both sides of the card.

vious example, 400 holes in the same area would then be required. Either these holes must be very small, or the component spacing must be increased. Otherwise, much of the printed-circuit track space will be wasted and many layers of interconnection will be needed. Thus, one of the driving forces for surface-mount technology is to decrease the size of the plated through holes which were previously quite large as determined by the diameter of the pins in PGAs or DIP. In these new component designs (PLCC, etc.), the leads are soldered directly to pads on the card or planar surface.

Memory packaging is the first area to benefit from this surface-mount innovation. As shown in Figure 4-11, both sides of the card planar can be used, and since the holes are small, extra wiring tracks can be included to handle the doubled component density. With memory, unlike logic, only one extra IO is required on the chip for each doubling of function. Thus, the small outline component can be made small to save space. Furthermore, the smaller component, because of the positioning of IO on all four sides and the low lead density, does not limit the spacing with a high wiring density demand. For logic, however, the situation is more complex. The benefits of the smaller component may be limited by an inability to wire the high IO count, even though significant wiring-track increases are possible. One drawback of surface mount, however, is that the surface pads themselves block wiring channels. As a result, in some cases an extra layer may be required for interconnections.

In any case, the common usage of 40 to 200 IO for surface-mount components (SMC), compared to DIPs that were previously limited to approximately 60 IO, is going to drive the printed-circuit industry to provide more complex composites. This is being accomplished by adding more tracks per inch within the individual layers or perhaps even increasing the use of multilayers. This technology is expanding because of the ever increasing use of circuit function at the chip level, spacing components more closely together for enhanced performance, and the continuing decrease in cost per circuit at the chip level.

As surface-mount technology expands into a broader range of components, including ceramics, and as the size of the ceramic components increases, more consideration is given to matching printed-circuit composites to the components in thermal expansion. Although there are several approaches to building matched composites, none have been produced in significant quantities. The simplest approach appears to be mechanical constraint of the laminate with power and voltage planes, where the proportions of the metals dictate the expansion properties as shown in Figure 4-12. When using this method, the voltage planes are constructed of trimetallic plate, copper-Invar®*-copper. The high strength and modulus of the metal then overwhelm the laminate expansion properties to the point at which they match ceramic or other componenets as desired [22]–[24]. The special procedure required for this method includes drilling the laminate, and cleaning and plating the innerplane contract areas. All other processes are the same as those in current use.

An alternative approach is to use quartz fabric [25], instead of glass fabric, in the laminate. In this situation, a thermal expansion match is possible without the presence of a continuous power or ground metal innerplane. Limited supplies of quartz fabric are available in the industry, but they are made of coarse yarns compared to those typically used in epoxy-glass laminates. Also, some drilling innovation is required because the increased hardness of the fabric causes greater attrition and breakage of the drill bits.

---

*Registered trademark of Carpenter Technology Corporation.

126 ADVANCED THERMOSET COMPOSITES

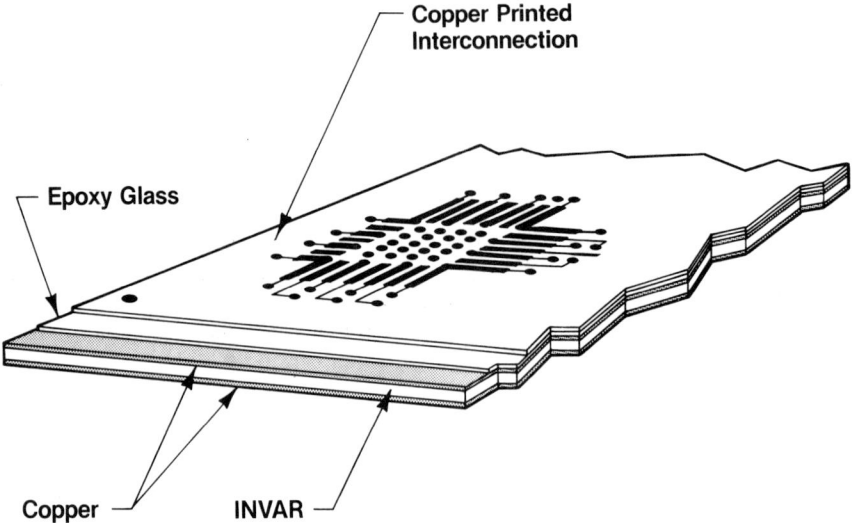

Figure 4-12. Matched expansivity composite positioned for small outline components.

Another method of matching in packaging hierarchy is done with Kevlar®* fabric, which can be used to create composites with extremely low thermal expansion [26]. In fact, these laminates have been used experimentally to replace ceramics by packaging chips directly [26]. These developments have been unsuccessful because of the nonadhesion between Kevlar fabric and the resin, and the difficulty of drilling Kevlar fabrics. It is possible that future innovations will overcome these problems.

Increasing the compliance or rubberiness of the surface has been another method of compensating for the hierarchical mismatch between expansion in the components and the printed-circuit composite [27]. This approach may, however, place excessive stresses on thin interconnections [17]. We conclude that surface-mount technology is still in its innovation stage, and the more stringent demands on interface-joining technology will delay rapid implementation.

While current usage is approximately 0.6%, the IPC [1] estimates that approximately 55% of the components will be SMCs by 1990.

Flexible Circuits

In the early 1980s, flexible cabling received a good deal of development attention, not only because of its ability to replace hardwired cables, but also because

---

*Registered trademark of E. I. du Pont de Nemours & Company, Inc.

the flexible cable can perform the functions of a printed-circuit composite, connector, and cable [28]. The base material of flexible cabling is usually polyimide that has been laminated to copper with an appropriate adhesive. The copper is eventually photoprinted into patterns. The flexible composite, typically a few thousandths of an inch thick including thin copper interconnections, is shaped as desired into various sizes of strips, which sometimes contain hundreds of interconnections. In contrast to flexible cabling, epoxy-glass composites are dimensionally stabilized by the glass cloth as well as the copper. The decreased stability of flex circuits allows more shifting and distortion, about .001 inch per inch, when the copper is removed by etching. It is possible to handle, solder, and join large networks together in highly compact, three-dimensional configurations as long as these dimensional changes are accounted for. A few examples are shown in Figure 4-13.

The flex circuits may be linked directly to epoxy-glass composites in strain-relieved configurations. If plated through hole joints are to be used, the flex cannot be laminated into epoxy-glass layered composites without a substantial stress resulting from the thermal expansion mismatch of the two materials.

Difficulties are also encountered in developing an adhesive system that will

Figure 4-13. Examples of flexible circuit applications tightly packaged. (*Courtesy:* E. I. du Pont de Nemours & Company, Inc.)

withstand all of the printed-circuit operations such as lamination, drilling, plating, cutting or slitting, and soldering.

In short, although flexible cabling technology is highly promising, the compromises, processes, and material interactions involved are not fully developed.

Nonetheless, the potential of very low cost processing and a reduction in assembly operation and hierarchy is resulting in wide usage of flexible circuits in one-sided and two-sided configurations. Western Electric Company, Inc. is an industry leader in the utilization of flexible cabling technology. They have developed the processing into a highly automated, continuous production line that turns out millions of square feet per year for communications applications. The U.S. market was worth $182 million in 1982 and is expected to double every two years [1].

DESIGN FACTORS

The main function of a printed-circuit composite is to provide conduction paths for the electronic components mounted on it. Also it facilitates heat sinking or thermal cooling of components through contact with structural elements in the electronic package. The electrical-function performance requirements distinguish the printed-circuit board from the other composite material systems. The necessity of putting electrical circuitry between dielectric layers and interlayer electrical connection also requires that printed-circuit manufacturing processes differ from other composite material systems. A set of design structures, normally considered important in determining density and performance, is shown in Figure 4-14.

The most important sets of dimensions are those that impact horizontal density like line width (w) and space (s), which determine—along with hole diameters (d)—the number of tracks or lines per grid site. Typical lines may be .005 to .01 inch, and holes may be .03 and .04 inch. State of the art technology may contain lines as narrow as 3.5 mils and holes with high aspect ratio as small as .016 inch. Line-to-line spacing considerations and line-to-plane spacings used are determined by compromises to be discussed, along with the technological capability.

Circuit Lines

The design of interconnection for printed-circuit composites is determined by a set of complex compromises, which begin with efforts to keep power low with high impedance (Z_0) devices. Therefore, 80 ohm impedance structures are often desirable. The next consideration is the width at which the interconnection lines may be etched with reasonable yields. These two factors combine to determine the spacing of the conductor with the referencing voltage plane. In the simple

ELECTRICAL AND ELECTRONIC APPLICATIONS 129

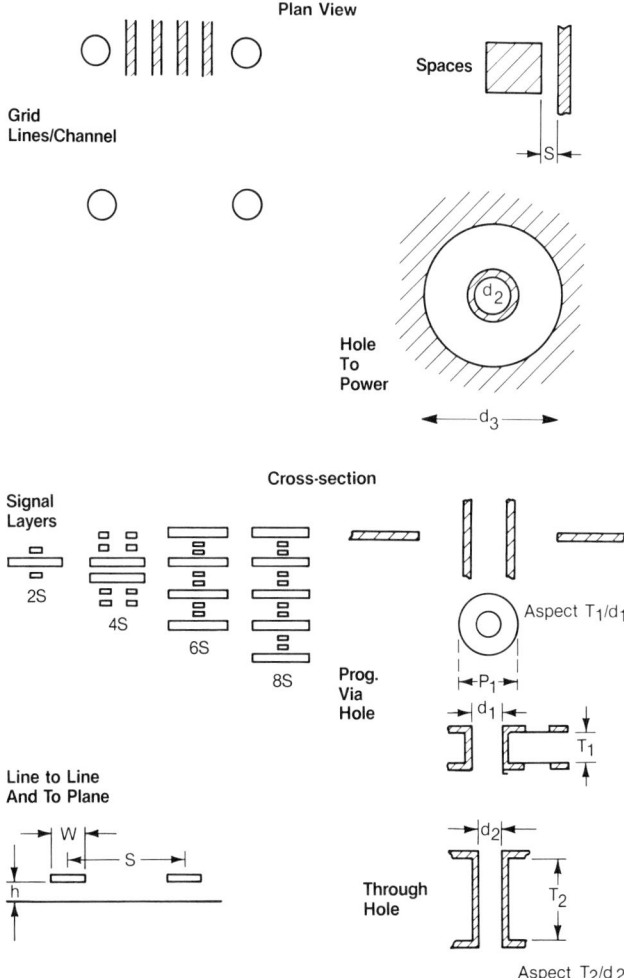

Figure 4-14. Design elements for printed-circuit boards.

closed-form approximation, the impedance is proportional to the half power of the inductance to capacitance ratio. This is easily calculated using free-space light velocity V_0 (3×10^{10} cm/sec), free-space capacitance C_0, and the relative dielectric constant ϵ_r. The resulting equation [29–31] for impedance is proportional to the reciprocal of the line width and proportional to the spacing with the referencing voltage plane. Figure 4-15 shows (a) a pair of signal lines between the voltage reference planes and (b) configurations having a pair of signal planes with reference voltage planes on only one side, along with typical di-

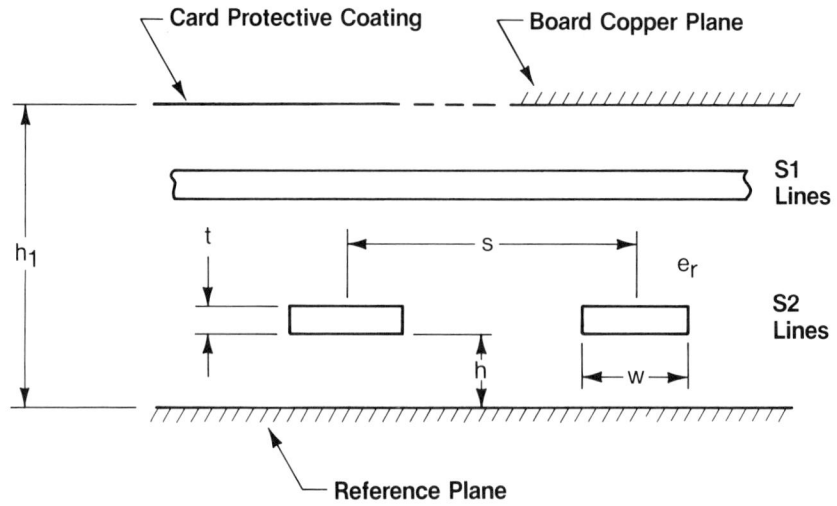

Figure 4-15. Card or board printed-circuit structure designed for impedance and electromagnetic coupling optimization.

mensions and the resulting impedance. The actual structure here and the more detailed design factors for cards and printed-circuit boards are discussed in detail by Arvanitakis et al. [31].

The spacing is very large between signal lines, 0.020 inch [31], in order to maintain very low electronic coupling in the structure with a reference voltage plane on only one side. With better shielding, the spacing can be decreased to

0.017 inch for reference planes on both sides. If the spacing to the ground plane can be reduced and the impedance maintained by using a narrower circuit line, the interconnection lines can be brought even closer together due to the improved shielding of the ground plane. Thus, the density of an 80 ohm structure is extremely dependent on extending the photoetching technology to its limits. The spacings of the dielectric materials are important, but not difficult to achieve.

In regard to the significance of holding the impedance tolerances, the objective is normally in the range of $+10\%$ to -10%. If these tolerances are not held, signal reflection [31] may occur, and the semiconductor device may not reach its threshold voltage for switching. Obviously, if the specification is indeed from $+10\%$ to -10%, deviation in either line width or line spacing to the reference plane must be less than this range. One can imagine that it is not at all easy to guarantee line widths of .004 to .005 inch to $\pm 10\%$ over large printed-circuit areas.

An approach to increasing density by decreasing the impact of electronic coupling involves utilization of wiring rules. For instance, a typical rule might allow closer spacing if the interconnection lines do not run parallel for more than a restricted length.

Another approach to increased density is decreased impedance. In this situation, the line is positioned proportionally closer to the ground plane. The increased shielding from the ground plane will allow the lines to be spaced more closely together for equal width at equal coupling. Since the spacing to the ground plane becomes very small for 50 ohm structures, special approaches to maintaining insulation may be needed to ensure the integrity of this spacing and to guarantee impedance tolerance.

The spacing is also dependent on the substrate dielectric constant, which is a function of resin and glass content. The use of epoxy-glass at a one-to-one ratio results in an ϵ_r of about 4.3.

To achieve the highest possible performance, a very low relative dielectric constant is desired [32–34]. Values of approximately 2.0 may be achieved with Teflon®* structures [34], while the value for polyimide in flex circuits without glass is about 3.5. The use of lower dielectric constant materials would allow proportionally closer spacing of signal lines at equal impedance. Thus, lower coupling levels would be achieved and signal line spacing could be decreased. To achieve the highest densities, diagonal wiring structures have also been considered [33]. The tracks in this case are at 45° angles to the accompanying signal pair, resulting in no significant electromagnetic coupling.

---

*Registered trademark of E. I. du Pont de Nemours & Company, Inc.

Plated Through Holes

With a 50 ohm structure, as compared to 80 ohms, the spacing to the ground plane is proportionally decreased. Therefore, using the 50 ohm structure will result in building a very thin printed-circuit board, thereby easing drilling requirements. Drilling difficulty increases with the requirements for high aspect ratio (length/diameter). The drill bending (and uncertainty in position by wander) increases exponentially with the ratio. Thus, the thickness of the composite—which depends both on the number of layers (including power and signal layers) and on the impedance—is an important factor in determining plated through hole design factors.

The stability of the composite materials is another factor in determining the space required for a plated through hole. The precision with which the drill is required to cut through the interconnection pad to a signal line is determined by the motion of the pad during previous lamination and chemical processes. Motion is determined by a variety of material and design factors. For example, if a solid copper plane is near the signal plane, the strength of the copper limits the motion. In the most precisely designed structures, this dimensional stability motion [35] is compensated for in the artwork. This may require artwork shifts, expansions, or typical contractions of about one part per thousand in some cases. The uncertainties in this required compensation factor can, however, be quite large. This uncertainty must be allowed for, by designing the connection pad for an inner plane large enough so that the hole will still be within its boundaries. Furthermore, extra spacing must be included in the insulation region between a plated through hole and the power plane. This is called clearance, which includes the dimensional stability factor, the drill wander, and artwork tolerances, as well as the insulation specifications. The predominant dimensional stability factor is that the laminate shrinks in the direction of the warp and expands in the fill direction during lamination due to residual stress relaxation of the glass cloth. Dimensional stability is discussed further in subsequent sections.

Programmable Vias

As mentioned earlier, holes are required to provide communication between signal planes as well as entrance of the component IO into the interconnection layers [36, 37]. The holes connecting signal planes need only to communicate between adjacent layers for the wiring versatility needed to move from X-direction tracks to Y-direction tracks. The position of these vias, therefore, is determined by the automatic wiring programs—hence, the name programmable. The holes are usually low aspect ratio. Therefore, they can be very small, limited only by the drilling factors: drill size, drill entrance and exit uncertainty, and a small dimensional stability factor. These holes may be mechanically or laser drilled. Since drill strength becomes very important for the smaller mechanical drills, it is expected that laser hole making will become more popular as the demand for small program vias increases.

Power Distribution

In addition to their significance in establishing an impedance reference system for the signal lines, voltage planes buried within the composite establish the voltage level for the circuits and the stability for these voltage levels. In the largest circuit boards, hundreds of amps may be switched at once. Thus, very large voltage drops can occur between the supply system through the power bus and through the voltage panels within the composite, causing large changes at the semiconductor components. The total switching voltage for the semiconductor, however, is in the range of a volt. Therefore, the voltage uncertainties allocated for the printed-circuit variations may be less than a hundred millivolts with approximatley equivalent shares for modules and circuits on the chips. The voltage planes minimize the voltage changes at the semiconductor component in two ways: by supplying the smallest possible resistance through a large area of copper at a reasonable thickness, usually .0014 inch, and by supplying a capacitance contribution when two planes are closely spaced in high performance circuit boards. It is advantageous, therefore, to build thick copper voltage cores that are very closely spaced in high performance circuit boards. In addition, large decoupling capacitor components are sometimes added for increased voltage stability.

In view of the complexity of the electrical structures, computer modeling of the total electrical response of a printed-circuit composite is necessary for an understanding of all the interactions. Several recent publications provide more information on this subject [31, 32].

COMPOSITE MATERIALS

The most commonly used substrate for high quality multilayer printed circuits, a laminate composite made of epoxy resin and glass cloth, is manufactured using processes that originated in the 1910s and 1920s. Although the processes are old, the epoxy resin was developed more recently, and the glass cloth industry was nonexistent until World War II. Both the glass cloth and the epoxy resins used in electric components have undergone substantial changes to accommodate increasingly specific and demanding electronic applications. The resins, which first achieved commercial significance in the 1950s, have grown substantially in use and variety [38].

A fundamental understanding of polymer structure and adhesion has existed since the 1930s. However, the processes by which these new materials were fabricated into completed printed-circuit boards were empirically controlled until the late 1960s, when aspects of the theory became integrated into the industrial processes. The driving force to employ fundamental concepts in control has been the rapid reduction in scale in electronics.

This section provides a brief description of the technology of printed-circuit laminate manufacture sufficient to reveal the areas of advancement and to indicate where progress is still needed. Applications of basic theory of control is one of the most rapidly progressing areas in the industry.

Glass Cloth

The process for making glass cloth involves drawing very fine filaments from the molten glass and combining as many as several hundred of these filaments into yarn [39]. The yarns are then woven into various kinds of cloth. The glass formulation initially developed to draw these filaments, C glass, had too many soluble, alkali-ionic constituents for good electrical performance. E (electrical) glass that is almost alkali ion free is now used for this purpose.

The original weights of glass cloth used in laminate production were manufactured for decorative applications. However, the drive toward thinner dielectric layers has resulted in the development of finer filaments and weaves as light as .001 inch thick. The final layer thickness in a laminate using this light weave would be slightly less than .002 inch thick.

Glass fabrics are described in several ways:

- Glass type includes differentiation among continuous and staple fiber, filament diameter, basic strand weight, the number of filaments in a strand, the number of strands twisted together, and the number of twisted strands plied together. A typical yarn designation is defined in Table 4-1(a).
- Strand designation appears in Table 4-1(b) along with the letter designation for filament diameter.
- Cloth or fabric style, which includes the number of yarns per inch, weave, thickness, etc., that go together to make a particular cloth style, is shown in Table 4-1(c).

Couplers

A solid lubricant called a size or sizing is applied as the filament is drawn to keep the fine glass filaments from abrading each other during cloth manufacture. Sizes, which were originally starch based, are now much modified and jealously

Table 4-1. Elements of Glass Fabrics

a. Typical glass yarn designation: ECD 900 1/0

    E = electrical grade (C = chemical, S = high strength)
    C = continuous filament (S = staple fiber)
    D = filament diameter. See under b.
    900 = ($\times$ 100—yd/lb). Thus 900 designates 90,000 yd/lb
    1/ = number of single strands twisted together
    /0 = number of twisted strands plied together. 0 indicates no plying.

b. Filament diameters. Filaments per strand

Filament Designation	Filament Diameter	Strand Count ($\times$ 100 = yd/lb)	Approximate Filaments per Strand
B	0.00015	900	204
		450	408
C	0.00018	150	816
		75	1632
D	0.00021	1800	51
		900	102

c. Typical fine industrial glass fabrics

Style	Construction Ends/Inch Warp $\times$ Fill	Thickness	Weight (oz/yd)	Yarns Warp	Fill	Weave
104	60 $\times$ 52	0.0012	.60	D900 1/0	D1800 1/0	Plain
108	60 $\times$ 48	0.002	1.45	D900 1/2	D900 1/2	Plain
116	60 $\times$ 58	0.0035	3.20	D450 1/2	D450 1/2	Plain

guarded compositions. They are still directed toward such aqueous operations as water-based dyeing. When glass cloth in the original sized or greige state is employed for prepreg manufacture, there is effectively little or no bond between polymer and glass. Various methods of sizing removal have been proposed. The most popular method is burning off the sizing in a direct flame or oxidizing the carbonaceous materials in a furnace. Such heat-cleaned cloth is used directly for a few composite applications, such as with silicone resins. More often, it is treated with a coupling agent or coupler that has two functions on one molecule—one reactive with the organic member of the composite and the other reactive with, or attracted to, the inorganic member.

Various types of chemical compounds have been employed as couplers [40]. One type that found early usage was the Volan®* couplers, chrome-organic complexes in which the chromium is associated with the glass surface of the cloth and the organic portion is associated with the polymer. Volan A, a chrome-

---

*Registered trademark of E. I. du Pont de Nemours & Company, Inc.

methacrylate complex, was used with epoxy resins. During the past 20 years, however, organosilanes have achieved almost complete dominance. Many different types of organic moiety are available for reaction with the polymeric material. The silane portion of the molecule is reactive with the glass or other inorganic surface and is frequently formulated as the trimethoxy- or triethoxysilane, as shown in Table 4-2.

Couplers are applied to the heat-cleaned glass from dilute aqueous, usually mildly acid, solution. The first chemical change that occurs during this process is hydrolysis, then partial polymerization, followed by a bonding of silicon to oxygen in the glass.

a) $H_2N(CH_2)_3Si(OCH_3)_3 \xrightarrow{H_2O} H_2N(CH_2)_3Si(OH)_3 \equiv R-Si(OH)_3$

b) $nR-Si(OH)_2 \longrightarrow HO-Si(R)(OH)-O-Si(R)(OH)-O-Si(R)(OH)-O-Si(R)(OH)-O-Si(R)(OH)-OH$ etc.

c) [diagram showing silane coupling to glass surface with $H_2N\ Cl^-$, NH_2, $(CH_2)_3$, Si, OH, and $(HCO_3)^-$ groups bonded through Si—O—Si linkages to the Glass Surface]

Dilute solutions are necessary for the coupling reaction because in more concentrated solutions the silanes polymerize excessively with each other and insufficiently with the glass surface. Furthermore, after they attach to the glass, they tend to build up thick layers with poor cohesive strength and little resistance to hydrolytic scission. Still another concern with the layers is clumping, which inhibits resin penetration of the glass yarn.

ELECTRICAL AND ELECTRONIC APPLICATIONS 137

Table 4-2. Chemical Structures of Common Couplers.

ORGANOREACTIVE GROUP	PURPOSE	TRADE NAME/ DESIGNATION
$CH_2=C(CH_3)-C(=O)\rightarrow CrCl_2$, $O-CrCl_2\leftarrow OH$ (methacrylate)	Polyesters, Epoxies	VOLAN A®
$CH_2=C(CH_3)-C(=O)-O(CH_2)_3 Si(OCH_3)_3$ (methacrylate)	Polyesters Polyparaxylylene	A-174[1]
$H_2N(CH_2)_3 Si(OC_2H_5)_3$ (amine)	Epoxies	A-1100[1]
$Cl(CH_2)_3 Si(OCH_3)_3$ (chloro)	Epoxies	A-143[1]
$CH_2=CH\ C_6H_4\ CH_2\ NH\ CH_2CH_2\ NH$ $Cl\ H\cdot(H_3CO)_3 Si(CH_2)_3$ (cationic styryl ammonium)	Epoxies	Z6032[2]

1. Manufactured by Union Carbide Corp.
2. Manufactured by Dow Corning Corp.

Organosilane couplers have been used for about 20 years to improve resin-glass adhesion; however, information about how they work was limited in detail. It was known that the silane portion oriented toward the glass and the organic portion toward the polymeric member of the composite. However, whether silanes merely associated with the glass or actually reacted with it was not established, and the selection of the organic part of the organosilane was Edisonian: couplers designed for methacrylates worked best for epoxies, and organosilanes with epoxy groups on them were inferior to many others for epoxy resins.

In large measure, this lack of understanding occurred because techniques to study interfaces were unavailable. Recently, a series of studies employing Fourier transform infrared spectroscopy detected a unique silane-oxygen-glass silica bond [41]. The ability to examine the direct chemical consequences of the employment of an adhesion promoter will inevitably enhance our capacity to selectively bond materials.

Epoxy Resin System

The resin system that is the workhouse in manufacturing both epoxy and glass cloth laminates for printed-circuit board applications is composed of a partially brominated, low molecular weight polymer based upon the diglycidyl ether of bisphenol A, to which at least 10% multifunctional epoxy resin has been added. A resin system based upon the polyglycidyl ether of a phenolic novolak is one example of this additive. While bromination imparts fire resistance, the multifunctional resin improves chemical resistance and raises the glass transition temperature (T_g). This resin is cured with about half the stoichiometric equivalence of dicyandiamide (DICY) and is catalyzed with benzyldimethylamine (BDMA), tetramethylbutanediamine (TMBDA), or other tertiary amines [42].

The virtues of this system include stability in prepreg form, reasonably good molding, production of a laminate with good physical properties, adequate interlaminar bonding strength, and a T_g of 120 to 130°C, depending on the formulation and process conditions.

Drawbacks include poor DICY solubility in epoxy resin solvents or resins, even at molding temperatures. Also, DICY is susceptible to moisture pickup, which impacts molding performance, reduces blister resistance, and impairs electrical properties. Further, a high T_g would make it easier to meet the heightened requirement level for today's printed-circuit boards, by reducing dimensional change and permitting greater control of flow during molding. Additionally, the solvents that are required to keep dicyandiamide in solution are toxic and they are harder to volatilize than the ketones usually used for epoxy resin solutions.

Even with the use of solvents such as dimethylformamide (DMF) or ethyleneglycol dimethyl ether (EGME), DICY frequently does not dissolve completely in the resin. It can form crystals in the prepreg, which at best changes the stoichiometry of the reaction. This, in turn, affects the flow during cure. These crystals may contribute to void formation in the laminate. Also, unreacted DICY, decomposed or not, may increase the propensity of the laminate to absorb moisture, with all the attendant difficulties.

While some of the detrimental features of the DICY-cured resins have only recently become significant, others have been evident since the early 1950s, when DICY was first utilized. Since then, a replacement curing agent has been sought, without success.

It is possible that the improved understanding of the chemical and physical changes involved in curing will indicate new methods to correct the evident faults of DICY. Careful analysis of the products of several compositions during a typical epoxy-amine cure [43] indicates that at least two, perhaps three, concurrent reactions are taking place. A series of investigations into the viscosity changes in thermosettings polymers [44–46] may improve the control of flow during lamination. New techniques for studying impregnation are being evaluated. As the results of these studies are applied to the technology of epoxy-glass laminates manufacture, it may be possible to overcome the undesirable features of the DICY-cured system. Certainly the wide usage that these epoxies continue to enjoy in the laminating industry demonstrates that their virtues are substantial.

Impregnation and Lamination

The manufacturing procedure for epoxy-glass laminates requires two major pieces of equipment, a treater and a high pressure laminating press. The treater, shown in Figure 4-16, is sometimes referred to as a coater. It is a continuous processor on one end of which glass cloth is unrolled, passed through a solution of the resin and hardener, and then through steel metering rolls to produce the right amount of resin add-on (pickup). The coated glass cloth next enters a forced air tunnel drier, which usually has at least two, sometimes four, heat zones. Diffusion to the surface removes the solvent, which evaporates while the resin is advanced to a smooth, dry coating on the glass cloth. The speed of the operation, 10 to 50 feet per minute, is determined by the diffusion process. If the polymer is reactive, as it is in epoxy-glass cloth prepregs, then the balance between drying and advancement of the degree of polymerization may dictate the aggressiveness of the drying conditions. Additionally, complications arise when two solvents with different volatilities are employed or when removal of the solvent precipitates the curing agent from the resin. This can occur, for example, when 4,4'-diaminodiphenylsulfone or dicyandiamide is used to cure epoxy resins. At the end of the drier, prepreg—the tack-free, resin-glass cloth combination—is rewound for storage before it is cut into sheets. At this stage, the resin is very brittle; without the glass cloth, it could not be handled effectively.

140 ADVANCED THERMOSET COMPOSITES

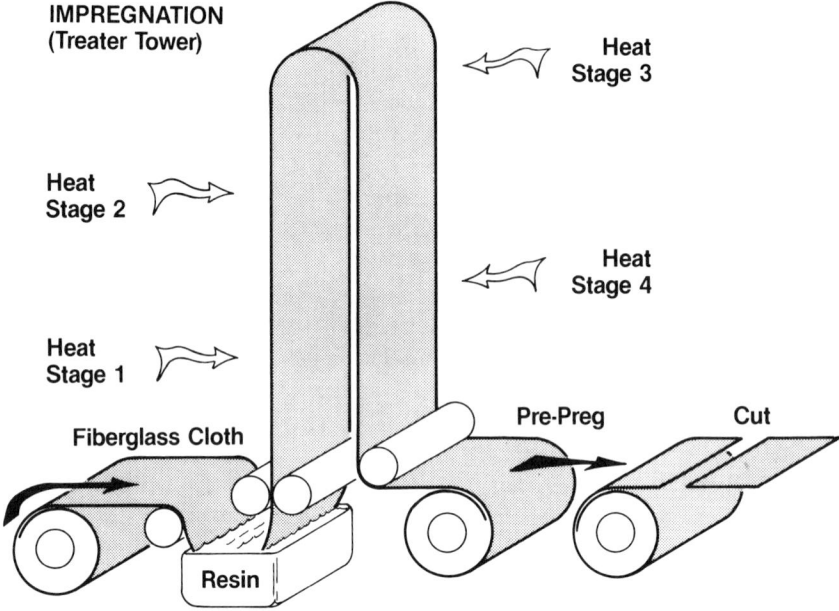

Figure 4-16. Treater for resin impregnation of glass cloth.

In the past, there have been few studies of desolvation of reactive systems like prepregs. Experimental prepreg treaters have customarily been used to produce small quantities of prepregs in resin chemistry studies, to carry out coupler evaluation, or to pursue material factors, rather than the drying/advancement process.

After being taken from storage, the sheets of prepreg are stacked (laid up) between polished, stainless steel planishing plates for placement between platens of the laminating press. Lubricant, separator film, or copper foil is placed between prepreg and the plates to prevent sticking; the copper foil is used in the production of copper-clad laminates.

A high pressure laminating press, as shown in Figure 4-17, can exert from 100 to several thousand psi pressure at temperatures up to several hundred degrees Celsius. Platen areas are typically as large as 1 meter by 1-1/2 meters, and the press can have as many as 16 to 20 platens. Both pressure increases and temperature profiles can be controlled in the heating and cooling portions of the prescribed cure cycle. In many instances, the cycle is selected to cure the resin completely. Many of the new high temperature polymer systems, however, must be cured at 200°C or higher to attain full property values. For these, a general purpose press, which rarely exceeds 180 to 190°C, is used to effect a partial cure at a convenient temperature. The cure is then completed by postbaking outside the press at the required higher temperature.

ELECTRICAL AND ELECTRONIC APPLICATIONS   141

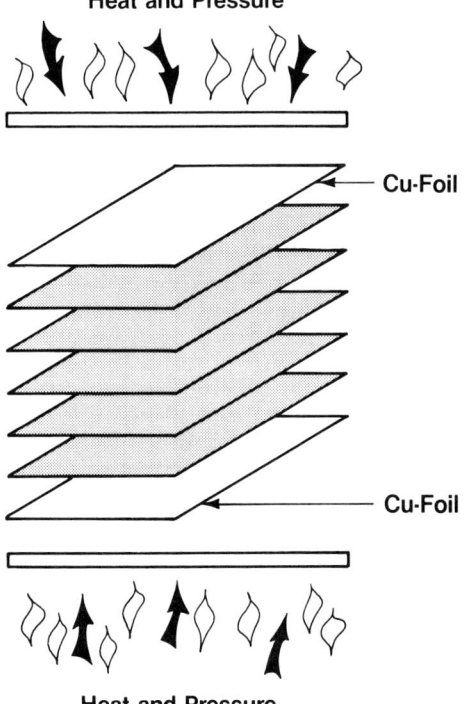

Figure 4-17. Lamination press.

Dimensional and Humidity Effects

Epoxy-glass laminate behavior through use of silane coupling agents is extremely important in producing high reliability laminates which may be used in uncontrolled environments. In fact, there are conditions of high humidty and voltage stress under which some couplers appear to lose their protective properties [47–50]. The consequence is the creation of shorting filament plated along the glass surface, which may penetrate from one circuit pattern to another, either hole to hole or line to hole. The accelerating factors for this phenomenon are humidity, voltage, and temperature. Thus, the major test of coupling agents is resistance to high humidity under voltage stress.

Once the resin is cured and strongly coupled to the glass through the silane coupler, the combined materials act as a composite with anisotropic mechanical properties. There is good evidence to suggest that residual stresses [51–54] are the predominant factors in affecting dimensional stability and its variations. These stresses arise from a variety of sources. One source is the glass fabric weave. In the fabric, the warp structure is very different from that of fill. During the impregnation process, tension is locked into the warp yarns of the prepreg which can be relaxed during the subsequent heating and lamination cycle when the resin becomes viscous. The flow and fill-in around the holes and dense circuitry areas, as well as the temperature variations across the laminate surface, give rise to localized stresses and variations of stresses. These and other process factors have been reviewed, discussed, and in some cases, measured [55, 56]. For example, ply orientation [57] is a major factor which can compensate for glass cloth tension. However, tracking these factors back through the process to a mathematical model is by no means straightforward. Although the literature discusses these theories at length, real data and detailed correlation are lacking. Most of the predictive models for dimensional changes are based upon some assumptions of deformations: orthotropic contraction, warp, twist, and other high order strain functions. In complex, multilayer board processing, the flowability of the partially cured resin-impregnated glass cloth, conditions of impregnation, lamination temperature and pressure, and dwell time are closely monitored and controlled to limit the variations of dimensional changes to specified tolerance limits. In some cases, highly accurate measurement techniques have been employed which, together with computer predictive models, assure that each individual layer of the composite will register to the others in a high volume production environment. Since changes in moisture and temperature expansion on the laminate layers are significant compared to the tolerance, they are usually temperature and humidity stabilized at critical processing steps.

Other Resin Systems

Resin systems with improved properties are being explored for printed-circuit laminate use. Superior T_g in epoxy systems can be achieved with increased cross-link density, which can be acccomplished by raising the amount of multifunctional epoxy in the formulation. This may mean that more epoxy novolak or an alternate formulation using, for example, the tetraglycidylamine of an aromatic diamine, is a possibility. Resins based on the latter material are used extensively as aerospace adhesives.

Condensation-type polyimides, like Kapton®,* are difficult to use in laminating because the water they give off during cure enhances blistering. The addition type has been employed for some period of time, either alone or in combination with other materials.

A hybrid resin formulated from a bismaleimide, a triazine polymer, and a tetrabromobisphenol A–base epoxy has a number of attractive properties. It employs common solvents such as methyl ethyl ketone, produces stable prepregs, is curable to high temperature usage laminates, and has excellent mois-

---

*Registered trademark of E. I. du Pont de Nemours & Company, Inc.

ture resistance and good electrical character. Other resins under consideration for aerospace applications are not yet used as electronic-composite contenders. The structure of these polymers is as follows:

New High Performance Polymers

CONDENSATION TYPE POLYIMIDES

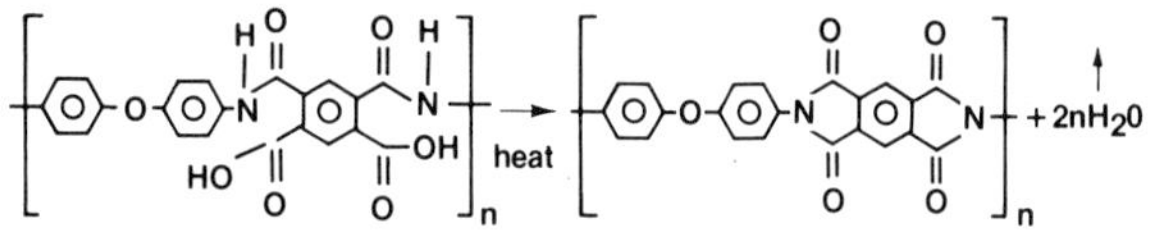

poly (amide acid) from oxydianiline (ODA) and pyromellitic dianhydride (PMDA)

polyimide form

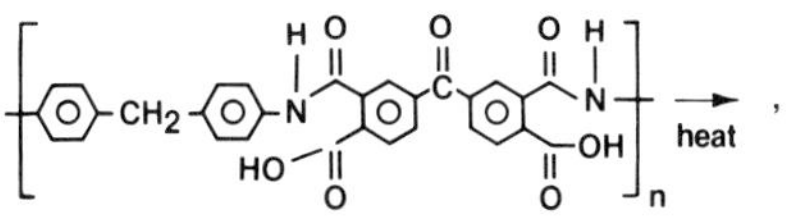

poly (amide acid) from methylene dianiline (MDA) and benzophenonetetracarboxylic acid dianhydride (BTDA)

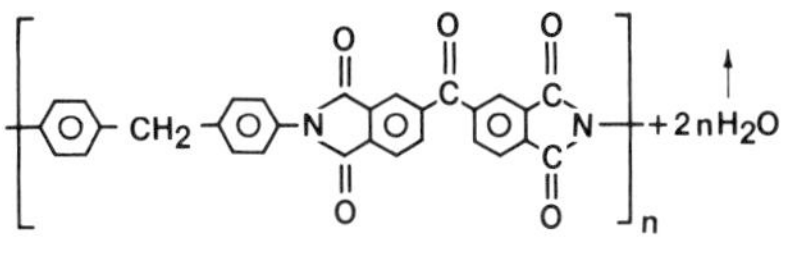

polyimide form

ADDITION-TYPE POLYIMIDE

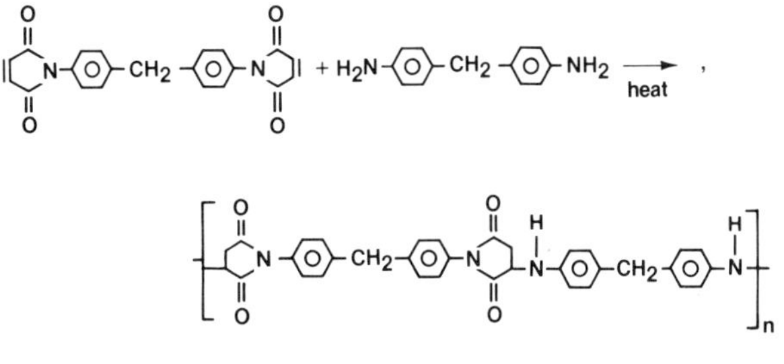

TETRAGLYCIDYLAROMATIC DIAMINE EPOXY RESIN

[Structural diagram of tetraglycidylaromatic diamine epoxy resin reacting with H_2NRNH_2 to form hydroxyl-containing crosslinked product]

BT Resin

[Structural diagrams of BT resin, bismaleimide-type structure, and isocyanate (NCO–Ar–C(CH₃)₂–Ar–OCN) trimerizing (~40%) to cyanurate structure with $R^1–OC$ and OR^1 groups]

[Structural diagram of tetrabromobisphenol A diglycidyl ether]

Polymerization takes place upon heating with zinc octanoate or other metal-organic catalyst. The resulting resin is a complex mixture of homo- and hetero-polymers.

Interfaces and Adhesion

Couplers used to improve adhesion between organic polymers and glass cloth were mentioned previously. Every other surface entering a multilayer printed-circuit laminate must also have some preparative treatment to enhance bonding to reduce interfacial defects. Particulates must be removed by rinses, sometimes with mechanical assistance. Grease and oil films prevent complete adhesion and must be displaced by either vapor or liquid solvent treatment. Even this kind of preparation may not provide an appropriate bond, particularly when the solid bonding surface is smooth.

It is extremely difficult to bond smooth polymer surfaces together. Therefore, roughening of some sort is customary. The most successful method is surface roughening effected by mechanical means. One popular approach is to mold the polymer against the rough bonding surface of copper foil, which is subsequently etched away. A chemical approach has been used for polyimides, by causing the surface material to revert to the polyamic acid stage [58], which makes it reactive and flowable in the presence of a supernatant uncured adhesive.

Smooth metal surfaces can be attacked by reagents that produce a rough, adherent compound permitting adhesion to the laminate. In this way, elevated temperature, alkaline oxidation of copper has been used to create a copper oxide layer that bonds well for moderate temperature application.

Alkaline Oxidation of Copper Surface for Bonding

$$ClO_2^- + 2H_2O + 4e^- \rightarrow Cl^- + 4OH^- \quad 0.76 \text{ V}$$

$$\underline{2[2Cu + 2OH^- \rightarrow Cu_2O + H_2O + 2e^- \quad 0.36 \text{ V}]}$$

$$ClO_2^- + 4Cu \rightarrow 2Cu_2O + Cl^-$$

One group of compounds used with some success on smooth metals is the chelates, such as benzotriazoles [61], which also act as excellent corrosion inhibitors. These compounds react with ionized metal at the metal surface, frequently the oxide, achieving association that is sometimes stable to above 200°C. Unlike organosilanes, the most successful of these chelates forms strong, coherent, multilayer films.

The adhesives used to join polymers to other polymers, metals, or ceramics and glasses, have become as specialized as the surface preparation. Some guidance is provided by the old organic rule for solvents: like dissolves like. Like also bonds like in many cases. Epoxies, for example, bond epoxies, but epoxies also bond thermoplastic and thermosetting polyesters, and many other polymers, to metals and to inorganic surfaces. Epoxies are very versatile adhesives, because they can be formulated with a wide range of reacting materials and are compatible with many fillers and modifiers with which they don't react. Polyurethanes are more versatile in mechanical applications, but they are inferior to epoxies electrically and do not possess the hydrolytic stability and temperature resistance of epoxies. As a consequence, polyurethanes are rarely used for printed-circuit bonding.

When the high temperature endurance of epoxy resin adhesives is exceeded, a few solutions can be obtained "off the shelf." For polyimide bonding, a family of high temperature acrylic adhesives is available (Pyrolux®*). For polytetrafluoroethylene (PTFE) bonding, only another perfluoropolymer is likely to provide sufficient thermal stability; a perfluoroethylene/perfluoropropylene (FEP) bonding film is effective if the PTFE surface is adequately prepared.

In the above applications, two solid surfaces are to be bonded. In one, unique, printed-circuit-like package, thin insulated wires are laid orthogonally in an embedment adhesive that has been coated on a structural substrate (Multiwire®†). This Photocircuit-formulated material must allow the wire to be embedded smoothly under ultrasonic heating by the dispensing head. It must immediately grip the wire to prevent displacement as the head advances or when another wire is laid across the first, and it must allow the wired pattern to be made flush under moderate pressure and temperature. Finally, it must be curable to fix the wire assembly. The material with these characteristics is a multiple-part composition of nitrile rubber, phenol-formaldehyde resin, epoxy, and chlorosulfonated rubber, plus curing system. Admittedly this is an extreme case. However, there are many equally demanding situations.

Flexible Circuit Materials

Most of the dielectric materials used in printed-circuit boards are semirigid composites, such as epoxy-glass cloth, epoxy paper, phenolic-paper, and polyester-fiberglass. However, flexible circuits constitute a rapidly growing fraction of the production for similar applications. Flexible circuits contain film materials such as polyethylene terephthalate, polysulfone, polytetrafluoroethylene, and polyimide. Either they are used as the only material separating the conductive layers, or more frequently, they are employed in conjunction with various adhesives.

Polyimides, along with polyethylene terephthalate and other polyesters, are the major dielectric materials used in flexible circuits (flex). For demanding applications, polyimides are the materials of choice; among these, the polycondensation polymers formed from oxydianiline (ODA) and pyromellitic dianhydride (PMDA) are the most popular. This is because they have the highest use temperatures among the commercially available film materials and because many modifications of the basic formulation are available. Resin systems for flexible circuits are shown on page 147.

Despite several drawbacks, such as high water absorption, susceptibility to attack by bases, and some difficulty in bonding when fully imidized, the polyimides have an attractive combination of properties. They are heat resistant to

---

*Registered trademark of E.I. du Pont de Nemours & Company, Inc.
†Registered trademark of Photocircuits Division of Kolmorgan Corp.

ELECTRICAL AND ELECTRONIC APPLICATIONS   147

COMMON NAME	FORMULA	TRADE NAME
Polyethyleneterephthalate	$\left[-CH_2CH_2OC\!-\!\!\bigcirc\!\!-\!C\!-\!O- \right]_n$ (with C=O groups)	MYLAR® [1] R DuPont
Polysulfone	$\left[-\bigcirc\!\!-\!\!\underset{CH_3}{\overset{CH_3}{C}}\!\!-\!\!\bigcirc\!\!-\!O\!-\!\bigcirc\!\!-\!S(=O)_2\!-\!\bigcirc\!\!-\!O- \right]_n$	UDEL® [2] R Union Carbide
Polytetrafluorethylene	$\left[-\underset{F}{\overset{F}{C}}\!-\!\underset{F}{\overset{F}{C}}- \right]_n$	TEFLON® [1] R DuPont
Polyimide (condensation type)	$\left[-\bigcirc\!\!-\!O\!-\!\bigcirc\!\!-\!N\!\!<\!\!\underset{C=O}{\overset{C=O}{}}\!\!\bigcirc\!\!\underset{C=O}{\overset{C=O}{}}\!\!>\!N- \right]_n$	
Polyimide (addition type)		Kerimid® [3] R Rhone-Poulenc

[structural formula for Kerimid addition-type polyimide with maleimide end groups, methylene-diphenyl linker, and diphenyl ether bridge]

---

[1] Registered trademark of E. I. du Pont de Nemours & Company, Inc.
[2] Registered trademark of Union Carbide Corp.
[3] Registered trademark of Rhone-Poulenc

148 ADVANCED THERMOSET COMPOSITES

350°C for brief excursions. They make strong films of good dielectric properties (ϵ_r about 3.2 and high dielectric strength), can adhere to prepared metal surfaces in the poly(amide acid) form, and can withstand many inorganic and organic processing chemicals. These features balance the greater cost of polyimides.

However, in the area of microwave applications, polyimides are not suitable. Microwave circuits demand that dielectric properties be matched to the particular use requirement. In this situation, the perfluorocarbon polymers, primarily polytetrafluoroethylene, are most suitable. Used frequently with glass fabric reinforcement, they are nevertheless uniformly difficult to process. They creep readily, drill with difficulty [59], and are extremely hard to bond. However, most significantly, they have flat capacitive response over many decades of frequency, are relatively uninfluenced by moisture or temperature change, and are capable of performing at very low dielectric loss levels [60].

Originally, flexible circuits were used as cabling and connectors, still the major volume application. However, as bonding methods improve, it may be possible to incorporate flexible layers into otherwise rigid laminates, provided the differences in expansivity can be compensated for in some way.

Fire Retardants

All organic materials are consumed by fire. A great many of them actually support combustion, including—among polymers—the epoxy resins. The degree to which a given organic polymer supports combustion in a composite depends upon its nature, how much of it is present, and what the other constituents are. A number of testing organizations, the most prominent being Underwriters' Laboratories (UL), have devised test methods to determine whether a particular composite structure is sufficiently fire resistant to minimize fire hazard.

In order to improve the fire retardancy to the point required by the appropriate test, a number of approaches are employed. They can be classed as: (1) compatible additives, (2) incompatible or insoluble additives, (3) structural modifications, and (4) combinations of the three.

A fire resistant material soluble in the resin can be added to the system. Chlorinated paraffins or phosphate esters (tricresyl phosphate or alkyl-ammonium phosphates) are examples. Such additives may not be soluble enough to impart the needed degree of fire inhibition, or if they are, they may be susceptible to migration out of the system. Further, they tend to change the physical or chemical nature of the system to which they are added, which may not be tolerable.

Insoluble materials like Dechlorane®* alumina hydrate or antimony trioxide

---

*Registered trademark of Hooker Chemical Corp.

are also used as flame retardants. They do not migrate, but getting a homogeneous distribution is not always possible, and fairly high levels of addition are required to attain appropriate levels of retardance. They are generally not used in epoxy resin glass cloth laminates, but find extensive employment in molding compounds.

Structural modification takes two routes. An element known to impart fire resistance may be chemically combined into the resin. Such elements are the halogens, for example. An instance of this approach is the use of brominated bisphenol A in epoxy resins. Polytetrafluoroethylene is another instance; PTFE is among the most fire resistant organic polymers. The second route involves determining what structures among the normal polymer compositions assist in fire retardancy by, for instance, charring instead of volatilizing. Some types of aromatic ring structures appear to be retardant in this way, and it is also known that higher degrees of cross-linking may contribute to fire resistance.

Combinations of any of the techniques cited may be employed and often are. Fire retardants formulated in this way frequently exhibit synergism, so that adequate retardancy is achieved at much lower levels in the sum than in any of the individual contributors alone. Because of this, and the difficulty of elucidating the nature of the fire retardant process, it seems that empirical solutions to meeting the test requirements for flame resistance will continue to be the common approach for some time to come.

Copper Foils

The conductive patterns in printed-circuit boards can be produced by additive (pattern-plating processes) or subtractive (etching processes). Each has its advantages. However, the majority of circuit patterns are made subtractively by etching photolithographic patterns from a copper foil laminate. Many metals have been made into foils, but copper foil presently accounts for almost all metal-foil-clad laminate production.

Printed-circuit quality copper foil is made by continuous plating process. The cathode consists of a slowly rotating stainless steel drum. Copper plates on the drum and is continuously stripped from it. Coil lengths are limited to a coil size that can be conveniently handled and can be 60 or more inches wide. Available common thicknesses range from .0007 to .005 inch, with the majority of applications using .0007 ($\frac{1}{2}$ ounce), .0014 (1 ounce), and .0028 inch (2 ounce), where ounce refers to weight per square foot.

The electrodeposition foil-making process has a number of advantages for electronics applications. Close tolerances can be maintained, and foreign materials and defects can be held to a very low production percentage. Additionally, the surface of the foil plated to the drum is smooth, facilitating precision etching. The surface toward the solution is normally rough and can be controlled by plating conditions. It is this rough dendritic surface that is used for mechanical bonding to the laminate. The adhesion can be further enhanced by chemical and electrochemical treatments [61]. Historically, the treatment was simply an oxidation. Now, however, other metals can be added to the roughened surface, and a variety of chemical transformations can be performed to bond with almost any specific resin or any use condition. Usually, the smooth side of the foil receives a thin coating to retard water staining. However, for applications where the surface is to be bonded to prepreg after a circuit pattern has been established, treatment similar to that applied to the rough side is given the smooth side as well.

PROCESSES

Process Sequence

In their simplest forms, printed-circuit composites are single and double sided. Figure 4-18 illustrates schematically two conventional methods, subtractive and additive, for manufacturing two-sided circuit composites (shown previously in Figure 4-3). For two-sided applications, copper is laminated on both sides of a base-insulating material. To facilitate interconnection between the two planes, holes are made where conductor paths will be positioned by the lithography, and the two planes are joined by plating the hole walls and surfaces. The circuitization step for the subtractive process is accomplished while the hole is protected, formerly by a liquid-applied resist and, more recently, by dry-film resist that provides a tenting over the hole during etching. In the alternative additive or pattern-plating approach, the resist is applied prior to plating, and the pattern and holes are plated at the same time. Then, the circuit pattern is protected by tin or solder while etching is accomplished.

For structures which are one sided and which do not require plated through holes, the patterns are etched prior to drilling or punching. At present, the holes are used only to hold components. With the double-sided board process, it is conceptually straightforward to proceed to the multilayer board step, as shown in Figure 4-5. The most common method is to produce double-sided laminates or cores, place partially cured prepreg between the cores to form a stack, and laminate them together into a single composite with two surface copper planes. The composite then is drilled to form the interconnection between the different planes, using either the subtractive or additive technique discussed earlier. In most multilayer boards, there are planes of solid copper in addition to the planes of circuitry. These provide power and reference voltage, as discussed on p. 4-30. While this increases the complexity of the multilayer board, the inclusion of the power planes often simplifies the total package performance and relia-

ELECTRICAL AND ELECTRONIC APPLICATIONS   151

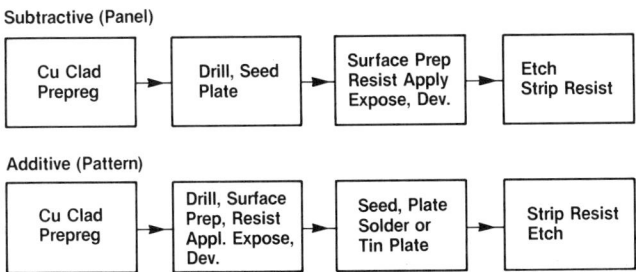

Figure 4-18. Basic circuitization processes: "subtractive etch" and "pattern plating."

bility. An alternative approach to composite building is a sequential buildup of layers. With the increase in circuit density on a plane and the number of multilayers, it is necessary to provide programmable vias (descrbed earlier) at the surface of the composite as well as between internal planes, to improve the wirability of the board. Figure 4-10 shows all of the layers and design features in perspective of a board built with sequential subassemblies [35].

Hole Making

The primary function of plated through holes in a printed-circuit board is to provide interconnection between two planes or from a component lead to a plane. Therefore, the holes must be positioned accurately through the different planes and must have copper deposited on the hole wall, with good adhesion to the glass-resin matrix and good metallurgical bonding to the exposed, internal copper lands of the interconnection plane. Moreover, to ensure insulation, the hole-making process should not produce fracture in the matrix material.

The most common hole formation technique for printed-circuit composites is mechanical drilling, although mechanical punching is still employed for thinner, nonreinforced substrates. Specialized multispindle drill machines [62] have been developed with high speed and feed to drill holes as small as .016 inch diameter. The drills are often back tapered and/or relieved to reduce heat generation from friction [63].

Drills are used commonly from 1/64 to 1/4 inch in diameter. For example, a commonly used PCB drill has a 0.040 inch diameter. Such a drill often has a four-facet point construction, 118° to 130° point angle, 12° to 20° primary relief angle, 30° secondary relief angle, and 30° to 50° overall helix angle. The drills are made from fine grain, tungsten carbide sintered-powder material with a cobalt matrix, usually 6 to 12%, which allow them to withstand the high stress and wear.

For high manufacturing throughput, high speed and feed are desirable. However, good hole quality requires proper selection of operating parameters such as spindle speed, feed and retraction rates, and dwell time. Representative operating ranges in the industry are: spindle speeds of 50,000 to 100,000 rpm, a feed of 150 inches per minute, and a retraction of 1000 inches per minute. The selection of drill design, materials, and construction, as well as the drilling parameters, is based upon a proper balance of many economical and technical factors. From the perspective of printed-circuit board hole quality, some of the important factors include drill breakage, drill wear, resin smear, resin-glass fiber fracture, hole entry/exit burr, nail heading (burr on an innerplane), and drill wander. Drill breakage is primarily caused by torsional and bending (buckling) stresses initiating a fracture around a weak region in the drill body. Increasing the cobalt content in the tungsten carbide drill can give it more resistance to torsion and buckling stress. However, increasing cobalt content reduces the material wear resistance and thus leads to duller drills for the same number of holes drilled (hits). Drill fracture usually initiates from a defect or void site. The carbide metallurgy and process strongly affect the drill performance [64]. It has been shown that defect or void occurrence is reduced with the high pressure isostatic process (HIP) and with smaller (1 to 2 micron), more uniform, carbide grain-size distribution. This results in improved drill flexure strength [65]. Microsurface finish has been reported to be effective in reducing drill friction and lengthening wear life [66].

Resin-glass fiber fracture is observed in the vicinity of plated through holes, even when there is good glass-resin adhesion. The fracture can be attributed to the pulling and pushing actions of the drill on the glass fibers in the heated resin during the cutting process. Therefore, it can be worsened by excessive spindle speed and high thrust.

Burring and nail heading, as well as hole roughness, are affected by drill point angle, drill advance per revolution, and drill sharpness. Entry and exit backer materials, which are usually made of phenolic or aluminum-clad composite, are widely used to reduce burring [67].

Drill smear, probably the most important factor determing hole quality, is caused by heating of the resin-chip debris and hole wall [68, 69]. In the repetitive drilling steps, the drill itself is gradually heated to a steady state, when the heat dissipation through the hole wall to the surrounding air and transport of the swarf are balanced by energy production from hole formation and surface rubbing. When the drill is new, much of this energy is carried away by the swarf. When the drill becomes dull, the frictional energy becomes dominant, and the resin in the swarf melts and settles around the copper innerplane surfaces. Eventually, the forces of thrust and frictional torque can fracture the drill. It is important to change the drill before the risk of drill breakage becomes significant. Weiss [68, 69] reported that, for a set of board and drill parameters, the temperature reached during drilling of a copper internal plane can reach 300°F in a functional land and 450°F in a nonfunctional land (which has less ability to conduct heat from the hole wall because it is not connected to internal circuitry).

For worn drills, Weiss estimated termperatures climb as high as 400°F in a functional land and 1200°F in a nonfunctional land. With a glass transition temperature of 265°F for epoxy resin, it is not surprising to observe epoxy resin smear over the copper innerplane and land, as well as land tear out, as reported by Weiss [68, 69].

The last step in preparing holes for plating involves cleaning of debris and contamination from innerplanes. Chemical etching with a variety of acids, bases and other chemicals is used in the industry [70]. More recently, interest has developed in using plasma for this cleaning step [71].

Laser Hole Making

The radiation from CO_2 lasers is absorbed by glass as well as epoxy. Thus the laser beams in the energy range of 30 to 100 watts can be steered and focused to provide a hole-making capability. In epoxy-glass, either the focusing or aperture technique can be used to make holes in the lower diameter range where mechanical drills are most difficult to use.

The implementation of the laser hole drilling into manufacturing was accomplished in 1979 and since that time has been used on a consistent basis to produce program vias joining the three interconnection plane pairs for the IBM 3081 Processor System printed-circuit board. There are approximately 9000 of these small vias (0.006 inch diameter) per board.

Photolithography

The starting point in the design of printed circuits is the layout, which designers develop from a schematic of the desired electronic circuit. Artwork masters of the circuit design are made on a scale of 1 to 1, 5 to 1, or 10 to 1, either by using rubylith precision tape on clear polyester film or by printing with a pattern generator from digitized information dirctly on a silver halide emulsion on a glass or polyester support. This pattern is then transferred to a master and subsequently to working photomasks through conventional photographic processes. Three types of photomasks are commonly used: silver halide emulsion, chromoxide emulsion, and diazo-sensitive coating on glass or polyester film. As shown earlier in Figure 4-18, these are used in the plating and etching sequences.

Single-sided, circuit-board fabrications use a copper, epoxy-glass laminate. This laminate is coated with photoresist, imaged with the desired circuit pattern, and developed, and the copper is etched away chemically, forming circuit lines. This photolithographic process for circuit-board fabrications is illustrated in Figure 4-19.

One of the most complex materials used in circuit-board fabrication is photoresist, a light-sensitive polymer coating which protects selected areas during chemical treatment [72]. Both liquid and dry-film photoresists are commonly used today. With negative-acting photoresist, the coating remains in the light-struck areas. Negative-acting photoresists characteristically have high chemical resistance and physical durability along with good image reproduction as well as high speed photoimaging capabilites. For these reasons, they are widely used in the manufacture of circuit boards.

Liquid Photoresist

Liquid photoresist contains at least two components, a film-forming resin and a solvent system. The resulting dried film undergoes polymerization when exposed to ultraviolet light. This chemical change increases the resistance of the coated, negative-acting resist to the developing solution, thus permitting dissolution of the material only in the unexposed areas. While it is possible for the resin itself to be intrinsically photosensitive, most resists contain sensitizers or photoinitiators, which absorb light and initiate a chemical change in the resin to produce an insoluble film. In addition to the resin, sensitizers, and solvent, various additives which enhance the performance of the resist, such as stabilizers, antioxidants, adhesion promoters, and coating aids, are present.

Common metal etch-type liquid resists, such as KMER® and KTFR®*, Waycoat IC and SC®†, and the Dynachem Micro G®‡, are based on derivatives of polyisoprene rubber. Figure 4-20 shows the chemical structure of rubber, which reacts with the bisazide sensitizer upon exposure to actinic (UV) radiation, to produce a tough, cross-linked material that is chemically resistant to metal etchants while adhering well to metal surfaces.

Uniform coating of the photoresist is critical since the thickness of this photoactive layer can impact subsequent processes. Precision spraying, roller coating, and dip coating are the most common coating methods used today for liquid photoresists. Each method has its unique advantages and limitations, such as equipment costs, precision, coating uniformity, and coating quality. Possibly the next most critical processes are exposure and development.

Uniform light exposure of he film is extremely important to obtain the desired line dimensions and the optimal physical and chemical properties. Inadequate exposure produces soft, poorly adherent patterns upon development (dissolution of the unexposed regions). The most widely used light source for negative-acting photoresists is the high pressure mercury-vapor lamp, which emits radiaiton in both the ultraviolet and the visual regions of the spectrum, 300 to 475 nm. This emission spectrum extends well over the region in which photoresists are most sensitive, as shown in Figure 4-21 and 4-22. The resulting line widths are a function of exposure, as illustrated in Figure 4-23. While optimum exposure produces line widths similar to the mask, it is possible to fine-tune line width with most photoresists by varying the exposure.

Unexposed areas on the photoresist film are subsequently redissolved in the solvent system from which the film was originally applied, thus producing a

---

*Registered trademarks of Eastman-Kodak Co.
†Registered trademarks of Philip A. Hunt Co.
‡Registered trademark of Dynachem Division of Morton-Thiokol Corp., Inc.

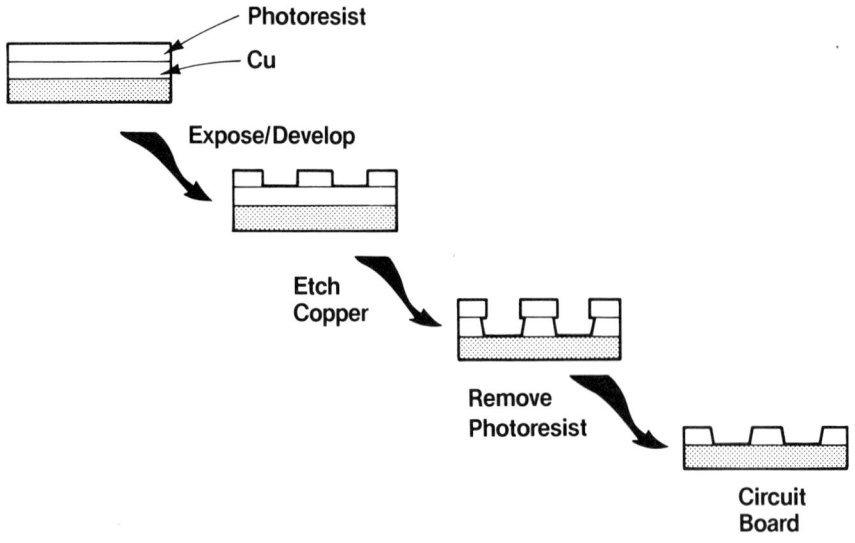

CROSS-SECTION — ILLUSTRATION

PROCESS	OPTIONS/FUNCTION
Resist Application	Spray, Dip, Roller-Coat
Air Dry	Bulk Solvent Removal
Prebake	Remove Entrapped Solvent, Harden Film
Expose	Define Desired Pattern, Cross-Link Light Struck Areas
Develop	Remove Resist in Non-Light Struck Areas
Rinse	Remove Residues
Etch	Remove Copper, Define Circuit Lines
Strip	Remove Remaining Photoresist

Figure 4-19. Negative-acting liquid photoresist processing sequence for circuit board.

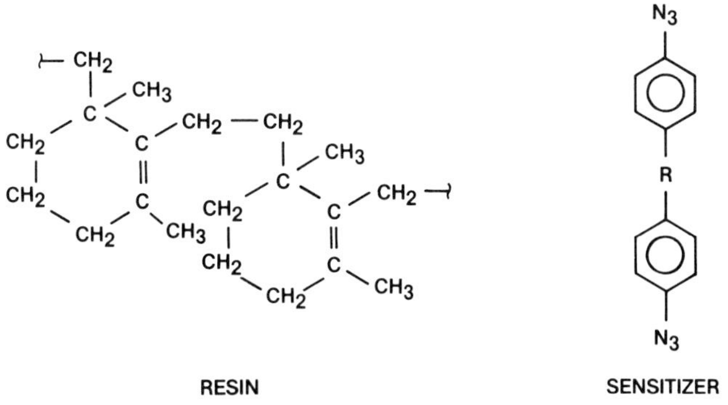

Figure 4-20. Polyisoprene rubber resin structure and bisazide sensitizer.

pattern or "windows" through which various chemical operations, such as etching, can take place. The most common development method involves spraying of the developing solvent onto the exposed/patterned photoresist, resulting in dissolution of the unexposed photoresist. Aliphatic or aromatic hydrocarbons, such as naphthas and xylenes, are common examples.

Dry-Film Photoresist

A large portion of photoresist applications for circuit-board fabrication utilizes dry-film-type systems. These, in their own right, are multilayer organic-composite systems comprised of a flexible photoresist film between release and cover sheets, as shown in Figure 4-24. During resist lamination, one cover sheet is removed from the photoresist and the resist is laminated to the panel through heat and pressure. The dry-film composite system offers advantages over liquid photoresists, since there is no solvent evaporation during coating and the surface is protected by a tough cover sheet during exposure and associated handling. Both sides of a two-sided circuit board can be coated simultaneously and the holes tented with photoresist. This hole tenting has been an important advance in the industry for protecting plated through holes during etching of the circuit pattern.

The dry-film photoresist process was described in 1968 by Dr. Jack Celeste of E. I. du Pont de Nemours, Inc. [73]. Use of dry film is more economical than wet systems from a total value viewpoint, including higher yields, elimination of complex coating equipment, and simplification of various processes.

158 ADVANCED THERMOSET COMPOSITES

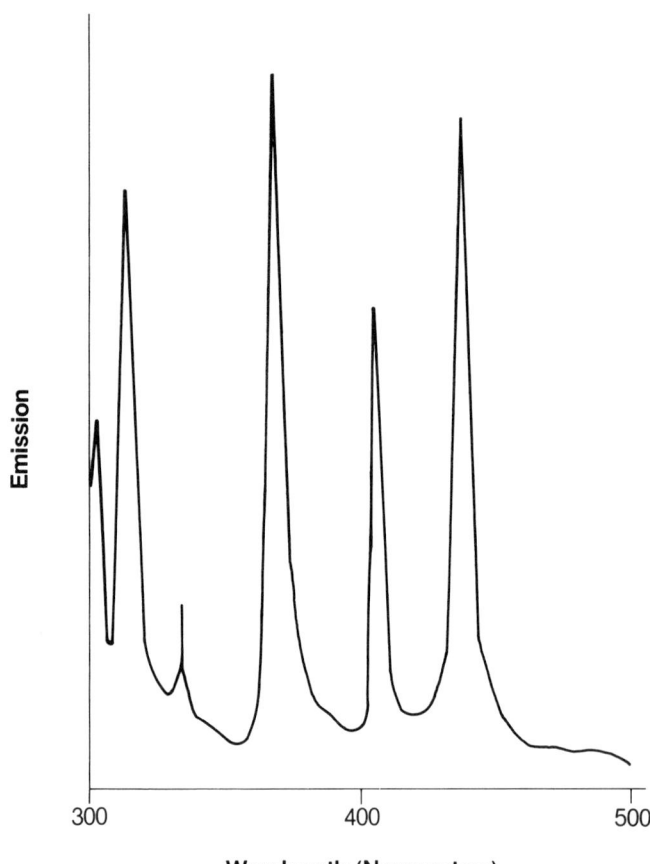

Figure 4-21. High pressure, mercury-vapor lamp emission.

One of the unique contributions of this composite system is the reduction of oxygen inhibition [74] during exposure, allowing more rapid and uniform crosslinking. In other words, the cover sheet serves a chemical as well as a mechanical function. The cover sheet does, however, have an adverse effect on resolution due to light absorption and scattering as well as forcing a separation of .001 inch between the photomask and the resist surface, thereby limiting the resolution.

On the other hand, an advantage of the dry-film systems is the various available resist thicknesses, which range from .0005 to .003 inch. This permits the use of dry film for electroless and electrolytic pattern plating. Newer dry-film systems, introduced in 1984, utilize a totally integrated application, exposure,

ELECTRICAL AND ELECTRONIC APPLICATIONS 159

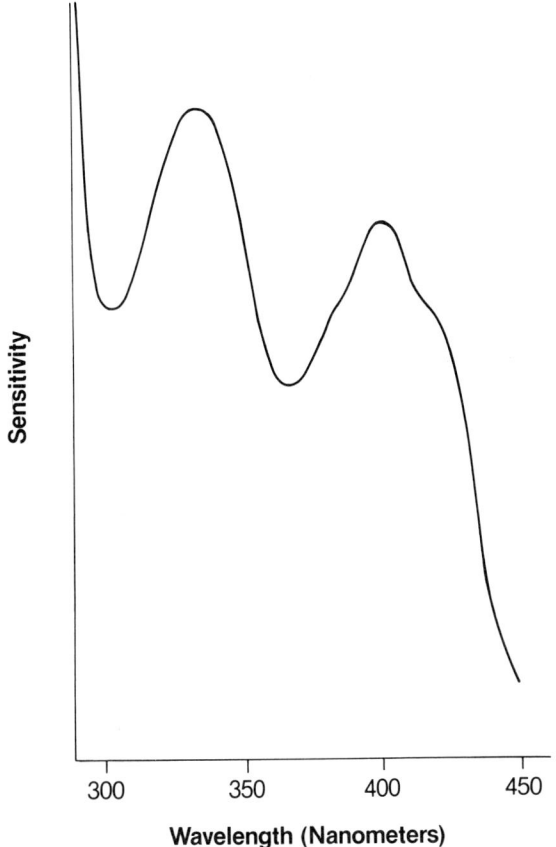

Figure 4-22. Typical negative-acting photoresist sensitivity.

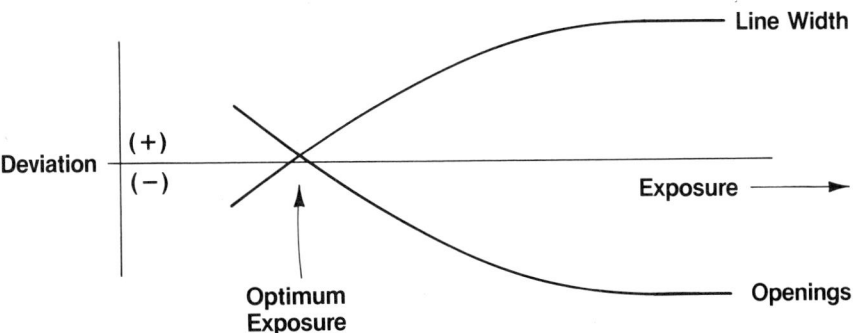

Figure 4-23. Pattern deviation as a function of photoresist exposure.

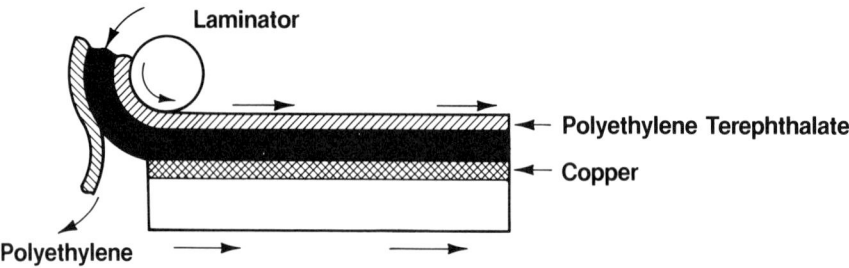

Figure 4-24. Dry-film photoresist system.

and development system, in which the cover sheet is removed prior to exposure, thereby improving resolution. The rapid exposure after cover sheet removal limits the time for oxygen diffusion into the film. Aqueous dry-film resists with semiaqueous, basic developers and strippers are being used with the new systems, eliminating or minimizing the organic solvent disposal concern. The aqueous developers [75, 76] pose fewer environmental concerns and are more economical.

Copper surface preparation has a significant impact on dry-film photoresist performance. Liquid photoresist has wide processing capability to wet, flow, and cover most copper surfaces. To achieve proper adhesion and intimate contact, dry-film photoresists are laminated onto the surface with flow achieved under heat and pressure. Mechanical scrubbing and chemical treatment are the most common techniques used today to help the adhesion. The surface treatments in order of preference [77, 78] are pumice scrub, nonwoven impregnated mat, and impregnated bristle. Adhesion promoters and conversion coatings are also used to increase photoresist endurance. This latter chemical treatment for adhesion can be a separate step, or the adhesion promotor may be incorporated in the photoresist.

All dry-film negative-acting photoresists in use today are based upon acrylate chemistry. The resists consist of a multifunctional monomer, such as trimethylolpropane triacrylate dissolved in a long chain poly(methyl methacrylate) binder. Additional chemicals are also incorporated for optimal performance, including photoinitiator, adhesion promoter, coloring dyes, latent-image leuco dyes, and plasticizers. Exposure of the film to ultraviolet light produces active, free radicals which initiate polymerization of the monomers through the photoinitiator and dye-functional oligomers. Development of the exposed panels is carried out in a spray developer. The quality of the resulting protoresist image as a result of developing is dependent on the chemical nature of the developer, temperature, spray droplet size, and energy.

Solder Masks

Final assembly of circuit boards usually requires a soldering step that uses a solder mask applied by screening, photoimaging, spraying, or dry-film laminate. This film becomes a permanent part of the composite. The dry-film solder mask, which has a process sequence similar to the dry-film photoresist just described, has many common processing advantages such as easy and convenient application as well as excellent photographic resolution. In addition, this material provides total encapsulation of circuit lines to avoid solder bridges. The permanent film provides environmental protection to the circuit lines. Similar to the dry-film photoresist, the dry-film solder masks consist of three separate layers—a polyolefin support, the solder mask photopolymer layer, and a polyester cover sheet (see Figure 4-24 for a comparison of these layers). These masks are also negative-acting films and are manufactured in various thickness up to .005 inch. Dry-film solder masks are suitable for covering high density circuitry with the constant thickness of an effective dielectric coating.

Screened Coatings

During early printed-circuit production, the simple patterns that were used allowed screened-on coatings, such as asphalt, to be used as etch resist. Today, screened coatings are used only as solder masks. A typical formulation [79] employs bisphenol A–based epoxy resins and, in addition, contains additives to control the flow during the screening process. Cure is accomplished through the reaction of liquid anhydrides with the epoxy group, thereby almost completely obviating the need for solvents. Modifiers and surface-action agents are frequently present, also. The virtue of the screened coating is low cost rather than pattern accuracy.

Copper Deposition

The development of the printed-circuit-board industry today owes much to the fact that high volume plating technology and processes have been available for efficient fabrication of circuits on nonmetallic surfaces. In a review paper, Van

Tilburg [80] traced the technical and commercial development of acid and alkaline copper plating. He indicated that the early baths used for printed-circuit boards were modifications of formulations for decorative purposes and failed the thermal stress tests for solderability. However, both acid and pyrophosphate baths have been developed specifically for printed-circuit applications, taking into account the need for high ductility copper. The electroless deposition was invented by A. Brenner and G. E. Riddell in 1946, and its early development has been recounted recently by Brenner [81] and by Pole-Baker [82]. Both the electrolytic and the electroless processes are used in printed-circuit fabrications, with the latter employed mostly in the additive technique discussed earlier [83]. The selection of the plating process for manufacture of circuit boards depends on requirements, such as line width, spacing, hole diameter, board thickness, and hole-aspect ratio, as well as materials compatibility. It is generally believed that the electrolytic processes are suitable for high volume, high throughput production without stringent requirements on hole-aspect ratio or line width. Electroless additive plating, however, with its capability for uniform deposition into recesses less accessible to the flow of current, would be suitable for high aspect ratio, multilayer boards with circuitry densely packed inhomogeneously.

In the electrolytic deposition process for printed-circuit boards, the pyrophosphate bath has been utilized well to provide good leveling power and good copper quality [80], principally through the use and control of additives [84, 85]. In recent years, much work in development and application has led to wide acceptance of acid-copper baths by many circuit-board manufacturers. The principal advantages have been improved plating speed, from 50 minutes per .001 inch to 25 minutes or less [86, 87] and high aspect ratio holes and fine line circuit [88, 89]. The improvement of acid baths, to a large extent, has been due to improvements in the quality of chemicals, in bath design, agitation, the introduction and control of additives [90, 91], and treatment for contaminants [92, 93].

Generally, all printed-circuit boards must have a copper deposit that sustains thermal shock at a temperature exceeding the melting point of solder. Tensile stress is induced in the soldering operation due to the differential thermal expansion between the laminate and the copper. This difference is very large in the plated through hole direction (Z-direction) due to the constraining effect of the glass cloth on the resin (X,Y-plane). If the induced strain is in the range of the ductility of the copper deposit, low cycle fatigue cracking could be induced from a few thermal shock cycles, particularly around the high stress concentration areas. Various tensile strengths and elongation-to-fracture data have been reported for samples plated in different copper baths. For thin copper films, the strength and ductility values could be affected by the nature of the test and test equipment. It is not easy to correlate strength and ductility data from one test technique to another. For example, in a recent paper [94], an additively plated copper deposit was tested at 3.1 to 4.5% true strain at fracture for bulge tests, and 4.3 to 7.9% at fracture for a uniaxial tensile test. Reverse bending is one way to measure low cycle fatigue resistance. With reverse bending used as a criterion, improvement in solder shock resistance for deposited copper has been reported [95]. A good test of plated copper reliability is testing the plated through holes based upon the thermal shock, simulating the soldering process. A study of the literature indicates that it is possible to have high quality, ductile copper by using the additive [96] or the electrolytic (acid copper or pyrophosphate) process [97]. There is sufficient variation among published literature data to indicate that the differences reported in copper quality are more related to stringent control and monitoring of the bath, rather than the inherent chemistry.

Reliability and Controls

Printed-circuit boards seldom if ever have a significant failure rate when they are produced with sufficient characterization and control; furthermore, any failure is usually readily analyzed and correlated to a process deficiency resulting in a defect or lack of integrity to specification which is easily recognized [98]. Most notable for concern are the resin and laminate defects which can give rise to loss of insulation integrity. However, the failure rates are so low that there is no good way of doing statistical studies with actual use data to correlate back to the process and contribute to improvements. The increasing demands for higher density and the migration of the electronic equipment into nonconventional environments (auto industry, homes) pose new concerns to establish improved materials and process controls for even better reliability. This can only be done by intense fundamental studies using improved characterization and test techniques.

EPILOGUE: THE FUTURE FOR PRINTED-CIRCUIT COMPOSITES

While the electronics equipment industry has been growing at a rate of 14 to 15% annually, the growth rate and uses for semiconductors have been explosive. A 1984 *New York Times* article quotes Robert J. McMillan [99], director of engineering at General Motors Corp. Delco Electronic Division, as saying, "There are at least five or six microprocessors in the average General Motors car, and, maybe a dozen in our luxury models." The same article shows that 120 million microprocessor chips are produced annually, with production doubling approximately every two or three years. Thus, at a rate of six microprocessor chips per printed circuit composite, usage can be calculated at approximately 20 million small composites, doubling every two to three years. The uses, as noted earlier, cover a broad range of products, including home computers, consumer electronics, and automotive applications as well as large computers.

The function in the microprocessor chips is also growing exponentially. Therefore, although the 1984 IO count for the units noted is in the range of 16 to 24 IO [100], future usage will extend from 64 to 240 IO. The high end of these units will be 32-bit microprocessors with 120 to 240 IO required for many applications. These powerful 32-bit units [34], capable of several million instructions per second, have increased in use from 5000 in 1983 to 20,000 in 1984. Memory chips with the same range of semiconductor devices, 64K to one million bits, will constitute the bulk of shipments made within the next few years. The consequences of this integrated circuit function growth will be an increased use of the small outline packages, tape-automated bond (TAB) [100] packages, and pin grid arrays [101], with average IO counts of 100 [100].

The drive for performance will continue relentlessly, with no foreseeable end. As noted earlier, the distance between components decreases as the reciprocal of the product of layers of wiring and tracks per channel on each layer. Thus, the most efficient package is one in which the chips, which are the limiting dimension, are packaged nearly edge to edge, as is the multilayer, multichip ceramic module or TCM [11]. One may, therefore, anticipate a thrust in which card and planar interconnection densities will approach the capabilities already demonstrated by multichip modules and the TCM used in the most powerful computer today. A common trend is in fact emerging [Table 4-3], and several preliminary examples have been demonstrated in the industry: chips backbonded and wirebonded directly onto printed-circuit boards [34], and leadless chip carriers attached to dense circuitry on insulated copper-Invar®-copper matched laminate [22, 102, 103].

The innovations which have brought us to the current state of circuit-board production are summarized in Table 4-3. New development trends require new materials, composite materials, mechanical and cooling structures, processes, and vitalization of new tools and techniques for control. These developments are resulting in increased emphasis on productivity enhancements and reliability improvements. The tremendous market growth of electronic materials, with one company already at $1 billion per year [8], is discussed in *Electronic Business 1984*.

To provide increased density and performance as well as tailoring to new application environments, a new family of composites is being developed to complement the epoxy, glass cloth, and copper foils which have served the industry so effectively in the last 20 years. Table 4-4, taken from a recent paper

Table 4-3. Most Significant Innovations in Multichip Packaging.

PROCESS TECHNOLOGY	FIRST-LEVEL PACKAGE	SECOND-LEVEL PACKAGE
JOINING	WIRE BOND TAB C-4	LEAD FRAME PIN-GRID ARRAY SURFACE MOUNT
INTERCONNECTION	DENSE THICK-FILM LINES DENSE PHOTOPRINTED LINES	DENSE PHOTOPRINTED COPPER LINES ETCHED & PATTERN PLATED COPPER
LAMINATION	LAMINATED CERAMIC MULTILAYERS	LAMINATED EPOXY GLASS LAMINATED COPPER-KAPTON
HOLE MAKING	PUNCHED LASER-DRILLING SCREENED VIAS	MECHANICAL DRILLING LASER DRILLING ELECTROLESS & ELECTROLYTIC PLATED VIAS & HOLES
E.C. & REWORK BY AUTOMATED WIRE BONDING	SINGLE WIRES	SINGLE WIRES & TWISTED PAIR
TOLERANCE CONTROL	DIMENSIONAL PREDICTION	DIMENSIONAL PREDICTION & ARTWORK COMPENSATION
ENVIRONMENTAL	PROTECTIVE COATING SEMI-HERMETIC SEALING HERMETIC SEALING	PROTECTIVE COATING

by Messner [104], shows some of the nonconventional laminate and foil materials available commercially today.

Many of these materials are not novel either in the broad composite technology industry or in electronic packaging applications. Today, packages can be found with polyimide/glass, polyimide/quartz [25], Teflon/glass [59], copper-Invar-copper, and injection-molded thermoplastic substrates [104]. Innovations

Table 4-4. Substrate Materials and Their Properties.

MATERIAL	RELATIVE DIELECTRIC CONSTANT	COEF. OF THERM. EXP. (PPM/°C)	THERMAL CONDUCTION W/M °C (Z DIR)	APPROX. COST FACTOR
EPOXY FIBERGLASS	4.5 — 5.0	14-18	.16	1X
POLYIMIDE FIBERGLASS	4.5 — 5.0	15-18	.38	2.5X
TEFLON FIBERGLASS	2.5 — 3.5	8	.25	15X
EPOXY KEVLAR	4.0 — 4.5	5.3 — 5.6	.12	3X
POLYIMIDE KEVLAR	3.5 — 3.6	5.6 — 5.8	.15	8X
EPOXY QUARTZ	3.6	5.0	.17	10X
POLYIMIDE QUARTZ	3.4	6-8	.20	14X
EPOXY GRAPHITE	—	3	1.0	—
THERMOPLASTICS	2.8 — 3.2	20	0.16	1.2X
PORCELAIN-COATED STEEL	6.8	12	1.0 25 (X-Y)	3X
COPPER-INVAR COPPER	—	6.4 — 5.8	16 165 (X-Y)	—
COPPER	—	17.3	450	3X
ALUMINA	9-10	6.5	20	6X
BERYLLIA	9-10	8.0	200	10X

such as coated metal substrate [24], shaped and flexible laminates, and many others will be examined, evaluated, and tested in the marketplace. The important thing is that the needs for these new applications are increasing rapidly, and the industry will have to gear up to meet the demand for rigid boards as well as flexible circuits.

The manufacturing processes suggested by Dixon will have more and more innovations in laser technology now being utilized in VLSI processes [105]. Productivity, automation, and continuous-process concepts for chemical processes will continue to receive further attention [106] to meet the expanding volume demand.

From the vantage point of the vacuum tube technology of 1944, it would have been impossible to predict the capability, even the nature of evolution, of today's printed-circuit composite technology (as described in "A History of Computer Packaging" by Werbizky and Haining [107]. It would likewise be presumptuous to try to predict the future direction 40 years from now. What is certain is that technology will be faster, more reliable, smaller, less expensive, and of a wider range of applications.

Composite technology is a highly demanding, interdisciplinary field—necessitating contributions from polymer scientists, chemists and chemical engineers, mechanics of materials, and process engineers. In electronic packaging composite technology, the need for such interdisciplinary science and engineering is evermore urgent. For example, the thrust toward reducing the size and frequency of defects in laminates requires much fundamental work on polymer kinetics and rheology. Fine line and small hole copper deposition processes will involve the cooperative efforts of surface chemists and electrochemists as well as photochemists, optical physicists, and polymer engineers. Metallurgists and mechanical engineers will be concerned with the ductility and strength of plated through holes and circuit lines, as well as with overall board structural integrity. Last but not least, the rapidly accelerating developments in surface and interface sciences, material characterization, and measurement techniques will have significant impact on the printed-circuit-board industry.

ACKNOWLEDGMENTS

The authors wish to thank C. L. Thompson, L. M. Coughlin, and P. Demetry for their editorial assistance.

REFERENCES

1. *The 1982 Market for Rigid Printed Wiring Boards and Related Materials. A Marketing Program for IPC*, and "The 1982 Market for Flexible Circuits." The In-

stitute for Interconnecting and Packaging Electronic Circuits, 3451 Church St., Evanston, IL 60203, May 1983.
2. "Market Forecasts," *Electronic Business*, **10**, No. 13, 188–190 (1984).
3. Lapin, P. J. "Printed Wiring Forecast," *Printed Circuit Fabrication*, 42–47 (February 1983).
4. Card, D., "Big, Bigger, Mature: Board Makers Grow Up," *Electronic Business*, **10**, 148–150 (1984).
5. Horton, J. and N. Compton, "Technological Trends in Automobile Science," **225**, No. 4662, 587 (1984).
6. Seraphim, D. P. "A New Set of Printed-Circuit Technologies for the IBM 3081 Processor Unit," *IBM J. Res. Develop.*, **26**, 37–44 (1982).
7. Bonner, R. F., J. A. Asselta, and F. W. Haining, "Advanced Printed-Circuit Board Design for High-Performance Computer Applications," *IBM J. Res. Develop.*, **26**, 297–305 (1982).
8. "Dupont Resist Materials and Image Systems," *Electoronic Business*, 86 (August 1, 1984).
9. "Mass Soldering of Circuit Packs," *The Engineer Journal—Western Electric*, 1 (1983).
10. Davidson E. E., "Electrical Design of a High Speed Computer Package," *IBM J. Res. Develop.*, **26**, 349–361 (1982).
11. Clark, B. T. and Y. M. Hill, "IBM Multi-Chip Multi-Layer Ceramic Modules for LSI Chips-Design for Performance and Density," *IEEE Transaction on CHMT*, 89–93 (1980). Blodget, A. J., Jr., "A Multi-Layer Ceramic, Multi-Chip Module," Proceedings of the Technical Program, 30th Electonic Components Conference, New York, 1980, pp. 283–285.
12. Werbizky, G. G., P. E. Winkler, and F. W. Haining, "Making 100,000 Circuits Fit Where at Most 6,000 Fit Before," *Electronics*, 109–114 (August 2, 1979).
13. Wilson, E. A., "Integral Liquid-Cooling System Simplifies Design of Densely Packaged Computer," *Electronics*, 123, (January 26, 1984).
14. Chen, W. T., L. C. Lee, C. K. Lim, and D. P. Seraphim, "Mechanical Modelling for Printed Circuit Boards," Proceedings of the Technical Program, Paper No. 41, Printed Circuit World Convention III, Washington, D. C., May 22–25, 1984.
15. Babuka, R., G. J. Saxenmeyer, Jr., and L. K. Schultz, "Development of Interconnection Technology for Large Scale Integrated Circuits," *IBM J. Res. Develop.*, **26**, 318–327 (1982).
16. Werbizky, G. G. and F. W. Haining, "New Bifurcated Spring Connector ZIF—System for the IBM 3081 Processor," Proceedings of the Technical Program, Productronica, Munich, West Germany, December 1981.
17. Balde, J. W. and D. Brown, "VLSI and the Substrate Connection, the Technological Tradeoffs of the Package Board Interface," A Multiclient Technical Review and Market Survey, done by D. Brown Associates, Inc., 1982.
18. Seraphim, D. P., "Chip to Module Interfaces," *IEEE Transactions on CHMT*, **1**, No. 3, 305–309 (1978).
19. Seraphim, D. P. and I. Feinberg, "Electronic Packaging Evolution in IBM," *IBM J. Res. Develop.*, **25**, 617–629 (1981).

20. Kawano, K., "Packaging Packs More Power into Mainframe Supercomputers," *JEE,* 63, (November 1983).
21. Schmidt, D. C., "A Model of the Impact of Integrated Circuits on Printed Wire Routing," Proceedings of the Technical Program, International Electronics Packaging Society, 1981, p. 143.
22. Reynolds, R. A., "Clad-Metal Core PC Boards Enhance Chip-Carrier Viability," *Electronic Daily News,* 211 (Aug. 23, 1984); "Clad Metal System and Components for Electronic Applications," TI Incorporated, Metallurgical Materials Division, Attenboro, Mass.
23. Wright, R. W., "Polymer Metal Substrates for Surface Mounted Devices," Proceedings of the Technical Program, National Electronic Packaging and Production Conference, Anaheim, Calif., March 1-3, 1983, p. 47.
24. Hauser, J., "Metal Core/Epoxy Glass Multilayer Boards for Reliability Interconnection of Leadless Chip Carriers," Proceedings of the Technical Program, National Electronic Packaging and Production Conference, Anaheim, Calif., March 1-3, 1983, p. 306.
25. Mahler, B., "Polymide/Quartz—A/C Chip Carrier Compatible Laminate Material," Proceedings of the Technical Program, National Electronic Packaging and Production Conference, Anaheim, Calif., March 1-3, 1983, p. 33.
26. Greer, S. E., "Low Expansion Organic Substrate for Flip-Chip Bonding," *IBM Technical Report* 2.2209 (July 10, 1978).
27. Fishman, D. and N. Cooper, "Mounting Leadless Chip Carrier onto Printed Circuit Cards," Proceedings of the Technical Program, IEPS, November 1981, p. 45.
28. Gunley, S., *Flexible Circuits Design and Applications,* Marcel Dekker, New York, NY, 1984.
29. Arvantakis, N. C. et al., "Coupled Noise Prediction in Printed Circuit Boards for a High Speed Computer System," Proceedings of the Technical Program, Seventh International Electronic Circuit Package Symposium, Los Angeles, Calif., 1966.
30. Harper, C. A., *Handbook of Electronic Packaging,* McGraw-Hill, New York, 1969.
31. Arvanitakis, N. C. and J. J. Zaora, "Design Considerations of Printed Circuit Transmission Lines for High Performance Circuits," Proceedings of the Technical Program, Wescon/81 Professional, San Francisco, Calif., September 15-17, 1981.
32. Davidson, E. E., "Electrical Design of a High Speed Computer Package," *IBM J. Res. Develop.,* **26,** No. 3 (1982).
33. Carlson, D. "High Performance Gate Array Packaging Design Tradeoffs," Proceedings of the Technical Program, IEEE Computer Packaging Spring Workshop, Split Rock, Pa., May 1982.
34. Beyers, J. W., E. R. Zellers, and S. D. Seccombe, "VSLI Technology Packs 32 Bit Computer System into a Small Package," *Hewlett Packard Journal,* **34,** No. 8, 3-6 (August 1983).
35. Bupp, J. R., L. N. Chellis, R. E. Ruane, and J. P. Wiley, "High-Density Board Fabrication Techniques," *IBM J. Res. Develop.,* **26,** 306-317 (1982).
36. Werbizky, G. C., and F. W. Haining, "Some Design Considerations Concerning

Blind and Buried Vias in Printed Circuit Boards," Proceedings of the Technical Program, International Electronics Packaging Society Conference, Baltimore, Md., October 29–31, 1984.
37. Sanker, N. G., "What Size Via—A Design Criteria," *Printed Circuit World*, III, Paper No. 60 (May 1984)
38. Markstein, H. W., "Laminates Support Technology Advances While Offering Alternate Choices," *Electronic Packaging and Production*, 83–89 (June 1984).
39. *Textile Fibers for Industry*, Owens Corning Fiberglass Corp. 1980.
40. Plueddman, E. P., *Silane Coupling Agents*, Plenum Press, New York, 1982.
41. Antoon, M. K., B. E. Zehner, and J. L. Koenig, "Spectroscopic Determination of the In-Situ Composition of Epoxy Matrices in Glass Fiber Reinforced Composites," *Polymer Composites*, **1**, 24 (1980). K. Ishida, et al., "Application of UV Resonance Raman Spectroscopy to the Detection of Monolayers of Silane Coupling Agent on Glass Surfaces," *Polymer Composites*, **2**, 75 (1981).
42. Chellis, L. N., U.S. Patent 3,523,037, "Epoxy Resin Composition Containing Bromianated Polyglycidyl Ether of Bisphenol A and a Polyglycidyl Ether of Brominated (Hydroxyphenyl) Ethane"(Aug. 4, 1970).
43. Byrne, C. A., G. L. Hagnauer, and N. S. Schneider,"Effects of Variation in Composition and Temperature on the Amine Cure of an Epoxy Resin Model System," *Polymer Composites*, **4**, 51–63 (1983).
44. Strecker, R. A. H., and D. M. French, "The Determination of Reactive-Group Functionality from Gel Point Measurements," *J. Appl. Polymer Sci.*, **12**, 1967 (1968).
45. Roller, M. B., "Characterization of the Time-Temperature-Viscosity Behavior of Curing B-Stage Epoxy Resin," *Polymer Eng. Sci.*, **15**, 405 (1975).
46. Macosko, C. W. and D. R. Miller, "A New Derivation of Average Molecular Weights of Nonlinear Polymers," *Macromolecules*, **9**, 199 (1976).
47. Lahti, J. N., R. H. Delaney, and J. N. Hines, "Characteristic Wear-Out Process in Expoxy/Glass PC for High Density Electronic Packaging," Proceedings of the Technical Program, I. Rel. Physics Symposium, 1979.
48. Lando, D. J., J. P. Mitchell, and T. L. Welsher, "Conductive Anodic Filaments in Reinforced Polymeric Dielectrics: Formation and Prevention," Proceedings of the Technical Program, I. Rel. Physics Symposium, 1979.
49. Tautscher, C. J., "Measuring Cleanliness of Printed Circuit Wiring—What Really Counts," Proceedings of the Technical Program, National Electronic Packaging and Production Conference, Anaheim, Calif., March 1–3, 1983, p. 556.
50. Marsh, L. L., R. J. Lasky, D. P. Seraphim, and G. S. Springer, "Moisture Solubility and Diffusion in Epoxy and E/G Composites," *IBM J. Res. Develop.*, **28**, 655–661 (1984).
51. Hahn, H. T. and M. J. Pagano, "Curing Stresses in Composite Laminates," *J. Composite Materials*, **9**, 91 (1975).
52. Hahn, H. T. and M. J. Pagano, "Curing Stresses in Composite Laminates," *J. Composite Mateials*, **10**, 91 (1976).
53. Marloff, R. H. and I. M. Daniel, "Three Dimensional Photoelastic Analysis of a Reinforced Composite Model," *Exp. Mechanics*, **9**, 156–162 (1969).

54. Chamis, C. C., "Lamination Residual Stresses in Cross-Plied Fiber Composites," Proceedings of the Technical Program, 26th Annual Conference of Society of the Plastics Industry Reinforced Plastics/Composites Division, 1971.
55. Daniel, I. M., T. Liber, and C. C. Chamis, "Measurement of Residual Stresses in Boron/Epoxy and Glass/Epoxy Laminates," Proceedings of the Technical Program, Composite Reliability ASTM STP 580, American Society for Testing and Materials, 1975, p. 340.
56. Daniels, I. M. and T. Liber, "Lamination Residual Strains and Stresses in Hybrid Laminates," Proceedings of the Technical Program, Composite Materials: Testing and Design (Fourth Conference) ASTM Special Technical Publication 617, American Society for Testing and Materials, 1977, p. 331.
57. Daniels, I. M. and Y. Westman, "Stresses Due to Environmental Conditioning of Cross Ply Graphite/Epoxy Laminates," Proceedings of the Technical Program, Fourth Internal Conference of Composition Materials, August 1980, Vol. 1, pp. 529–542.
58. Linde, H. G. and R. T. Gleason, "Adhesive Interface Interaction between Primary Aliphatic Amine Surface Conditioners and Polyamic Acid/Polyimide Resins," *J. Polymer Science*, **20,** 1031–1041 (1982); "Thermal Stability of the Silica Aminopropylsilane-Polyimide Interface," *J. Polymer Science*, **22** (1984).
59. Smith, T. A., "Drilling and Plating of Teflon/Glass Printed Circuit Boards," Proceedings of the Technical Program, National Electronic Packaging and Production Conference, Anaheim, Calif., March 1–3, 1983, p. 293.
60. "3M Sourcebook, Electronic Products Div," Publication No. 225-4S, 3M Center, St. Paul, MN 55144.
61. Poling, G. W., "Inhibition of the Corrosion of Copper and Its Alloys," Final Report INCRA Proj. No. 185, February 1979.
62. Wrenner, W. R., "Large Multi-layer Panel Drilling System," *IBM J. Res. Develop.*, **27,** 285–291 (1983).
63. Nelling, J. M., R. W. Saxton, and W. R. Hewitt, "Micro Drill Technology Update," *PC Fabrication* (March 1984)
64. Tuck, J., "Small Hole Drilling," *Circuit Manufacturing*, **23,** No. 2, 24–32 (1983).
65. Schaefer, R. T. and R. J. Houseman, "Carbide Metallurigcal and Engineering Applied to PCB Drill Bits Printed Circuit Fabriaction," July 1983, pp. 45–55; October 1983, pp. 14–18.
66. "Drill Parameter Study," Hughes Aircraft Company, Res. and Dev. Report COR-ADCOM-772640-F, 1979.
67. Deitz, R., "Tune Up Your Drill Operation," *Printed Circuit Fabrication*, 33–37, (April 1983)
68. Weiss, R. E., "Evaluation of Drilled Hole Quality as a Function of Speed and Feed in Mutilayer Boards," *Circuit World*, **7,** No. 1 (1980).
69. Weiss, R. E., "The Effect of Drilling Temperature and Multilayer Board Hole Quality," *Circuit World*, **4,** No. 3 (1977).
70. Deckert, C. A., E. C. Couble, and W. F. Bonetti, "Improved Post-desmear Process for Multilayer Boards," *Printed Circuit World*, III (May 1984).

71. Rust, R. D., R. J. Rhodes, and A. A. Parker, "The Road to Uniform Plasma Etching of Printed Circuit Boards," Proceedings of the Technical Program, Printed Circuit World III, Washington, D. C., May 22–25, 1984, Paper No. 42.
72. DeForest, W. S., *Photoresist Materials and Processes,* McGraw-Hill, New York, 1975. Textbook on basic photoresist processes, applications, and technology.
73. Celest, J. R., Initial Patents on Dry-Film Photoresist, U.S. Patents 3,448,089; 3,458,311; 3,475,171; and 3,469,982 (1964).
74. Barnes, C. E., "Oxygen Inhibition," *J. Am. Chem. Soc.,* **67,** 217 (1945).
75. Fullwood, L., "Factors Affecting Fine Line Reproduction in Dry-Film Photoresist," *Insulation/Circuits,* 47 (January 1982).
76. Crymes, C., "Totally Aqueous Dry-Film Comes of Age," *Printed Circuit Fabrication,* 68 (January 1984).
77. Hamilton, W. L., "Dry Film Performance on Electroless Copper," *Electronic Packaging and Production,* 232 (January 1981). Clampitt, J. H., "Getting the Best Dry Film Performance," *Printed Circuit Fabrication,* 35 (June 1984).
78. Neal, J. S., "Mechanical Surface Preparation," *Printed Circuit Fabrication,* 61 (June 1983).
79. Kenney, E., D. Lazzarini, and R. Winters, U.S. Patent 4,292,230. September 29, 1981.
80. Van Tilburg, G. C., "75 Years of Copper Plating," *Plating and Surface Finishing,* 78–82 (June 1984).
81. Brenner, A., "Reminiscences of Early Electroless Plating," *Plating and Surface Finishing,* 24–27 (July 1984).
82. Baker-Pole, M., "Printed Circuits—Origin and Development, Part 2," *Circuit World,* **10,** No. 3, 8–14 (1984).
83. Alpaugh, W. A. and J. M. McCreary, "Copper Plating Advanced Multilayer Boards," *Insulation/Circuit,* 24–27 (March 1978).
84. Janitz, M., C. Ogden, D. Teuch, and R. Thompson, "Operation of Copper Pyrophosphate Circuit Board Plating Bath at High Additive Concentrations," *Plating and Surface Finishing,* 58–60 (January 1984).
85. Lowry, B., C. Ogden, D. Teuch, and R. Young, "Production Implementations of Controls for Copper Pyrophosphate Circuit Board Plating Baths," *Plating and Surface Finishing,* 70–74 (September 1984).
86. Cavano, M., "High Speed Copper Plating for Printed Wiring Board," *Printed Circuit Fabrication,* 33–35 (July 1983).
87. Gornell, J., "High Speed Copper Plating," *Printed Circuit Fabrication,* 73–74 (July 1984).
88. Smith, R. E., "Electrolytic Plating for High Aspect Ratio Holes and Fine Line Circuitry," *Printed Circuit Fabrication,* 66–67 (April 1983).
89. Peterson, G., "Precision Plated-Through Holes," *Printed Circuit Fabrication,* 50–58 (July 1984).
90. Freitag, W., C. Ogden, D. Teuch, and J. White, "Determination of the Individual Additive Components in Acid Copper Plating Baths," *Plating and Surface Finishing,* 55–60 (October 1983).
91. Naneu, N., L. Mirkova, and S. T. Rashkov, *Surface Technology,* **11,** 117–127 (1980).

92. Anderson, D., M. Hanna, C. Ogden, and D. Teuch, "Purification of Acid Copper Plating Baths," *Plating and Surface Finishing*, 70–71 (December 1983).
93. Fritz, D., and B. Sullivan, "Copper and Solder Electroplating Contaminants," *Printed Circuit Fabrication*, 36–43 (July 1983).
94. Parente, M. and R. Weil, "Tensile Testing of Electrodeposits," *Plating and Surface Finishing*, 114–117 (May 1984).
95. Houma, H. and S. Mizushima, "Application of Ductile Electroless Copper Deposition on Printed Circuit Boards," *Metal Finishing*, 47–52 (January 1984).
96. DeBritta, J. F., "High Reliability Electroless Copper for Fully Additive PWB Manufacture," *Printed Circuit World*, III, No. 69 (May 1984).
97. Horbay, P. D., "Reliability Study of Copper Plating," *Printed Circuit World*, III, No. 45 (May 1984).
98. Smith, G. A., "A Look into Future Amendments and Revising of Commercial MIL Specs," *Printed Circuit Fabrication*, 40–51 (November 1983).
99. Sanger, D. Z., "The Great War over Superchips," *The New York Times*, Section 3, pp. 1, 8 (September 9, 1984).
100. Marshall, J. F., "New Applications of Tape Bonding for High Lead Count Devices," *Solid State Technology*, 175 (August 1984).
101. Knebusch, M., "Shrinking Circuit Boards with Surface-Mount Components," *Machine Design*, 69 (May 24, 1984).
102. Inpyn, B. E., "LamCore MLB's Reliable LCC Interconnections," *Society for Application Materials and Process Engineering Quarterly*, 3 (April 1984).
103. Marksstein, H. W., *Electronic Packaging and Production*, 50 (April 1984).
104. Messner, G, "Nonconventional Substrates," *PC World*, III, No. 34 (May 1984).
105. Dixon, T., "Laser Brings Precision to Electronic Manufacturing," *Electronic Packaging and Production*, 98–108 (March 1984).
106. Eidschun, C. D., "Process Automation for Printed Circuit Manufacturing," *Printed Circuit Fabrication*, 78–83 (February 1984).
107. Werbizky, G. W. and F. Haining, "A History of Computer Packaging," *IBM Technical Report* 2916 (October 22, 1984).

5
AUTOMOTIVE APPLICATIONS OF COMPOSITES

Peter Beardmore
Metallurgy Department
Research Staff
Ford Motor Company
Dearborn, Michigan

The automotive industry has long been considered to be very traditional in its use of both materials and technology. In the last decade, a quiet revolution has been taking place in the transportation industry, and it would be more accurate today to refer to it as a high technology industry. For example, the automobile industry is the largest single user of custom designed integrated circuits, which are used for controlling engine functions, and in the future, the number of these microprocessors on the vehicle will increase dramatically. In line with this high tech computer usage, materials have been undergoing, and will continue to undergo, a major change in many applications throughout the vehicle. The major driving force for these rapid changes is of course, a striving for increased efficiency, reliability, and customer appeal.

The materials that are the most likely candidates for extensive substitution are high strength steels (HSS), aluminum alloys, and a wide variety of plastics and reinforced plastics (composites). The current trends in these materials are shown in Table 5-1, which compares usage in 1980 with projections for 1985 and 1990. Based upon these current trends, it can be projected that the average vehicle in 1990 will utilize about 380 pounds of HSS steel (i.e., approximately 16% of vehicle dry weight), about 260 pounds of plastic (~11% of vehicle weight), and about 140 pounds of aluminum (~6% of vehicle weight).

In the nearer term, high strength steels represent probably the most cost-

AUTOMOTIVE APPLICATIONS OF COMPOSITES

Table 5-1. Average Materials Usage (Pounds) in Automobiles in the United States.

Material	1980	1985	1990
HSS	250	270	380
Aluminum	120	135	140
Plastics	220	245	260

effective means to achieve weight reduction. Depending upon the class of steels used and the particular application, between 10 and 30% weight reduction can be achieved with HSS steels. Aluminum clearly offers more substantial weight reduction opportunities, on the order of 50%, but frequently involves a significant cost penalty. This is inevitably true of wrought aluminum products, and little additional usage of this form of aluminum is anticipated in the future. Cast aluminum components can, however, compete economically and will continue to hold a place in the automobiles of the future.

The most dramatic growth of all automotive materials has occurred in plastics. Figure 5-1 shows the growth in plastics usage since the 1940s and demonstrates the precipitous increase occurring throughout the 1970s and early

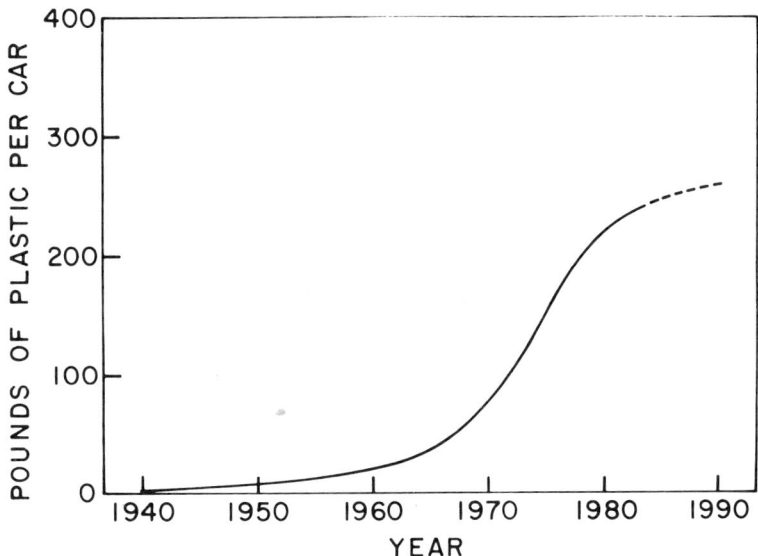

Figure 5-1. Growth in plastics usage in automobiles since 1940.

Table 5-2. Newer Plastic Materials.

- Reinforced urethanes
- High heat distortion thermoplastics
- High glass loaded polyesters
- Structural foams
- Super-tough nylons
- High molecular weight polyethylenes
- High impact polypropylenes
- Polycarbonate blends

Table 5-3. Developing Plastic Processes.

- Reaction injection molding of urethanes
- Compression molding of SMC
- Structural foam molding
- Blow molding
- Thermoplastic stamping
- Resin transfer molding

1980s. Tables 5-2 and 5-3 list some of the new materials and new manufacturing process developments, which have been a key part of the accelerating cost-effective use of plastics in vehicle applications. Except for isolated exceptions, this rapid increase in plastics has, however, been confined to decorative and semistructural (at best) applications within the automobile. If plastics are to achieve a projected utilization greater than the 260 pounds indicated in Table 5-1, applications in the structural load bearing components must be attained. Radiator supports, drive shafts, frame cross-members, leaf springs, and bumpers are typical applications that are current by being aggressively explored to capitalize on the 50% or more weight reduction opportunities exhibited by many plastic composite structures. High performance, fiber reinforced composites obviously offer the opportunity to withstand the load environment in the structural parts of a vehicle. Any such applications would have a major impact on the consumer marketplace. Over 600 million pounds of glass fiber reinforced plastics were used by the transportation industry in the 1980s. It has been projected that this amount per year will exceed one billion pounds per year over the next five years. Analysis based on expectations of technolgical and, in particular, manufacturing developments suggests that fiber reinforced polymer composites have the distinct potential to become a real force in the next-generation class of materials for vehicle applications. The competitive nature of the automobile business will ensure that all the relevant factors (material costs, investments, weight savings, manufacturing considerations, production rates, etc.) are carefully developed and optimized before a final verdict on the degree of usefulness of these high performance composites in the vehicle of the future can be made. The opportunity for these high performance materials to make the transition from expensive aerospace-type applications to consumer oriented, everyday product applications has never been greater.

MATERIALS AND DESIGN CRITERIA

Unquestionably, the fiber which will find the greatest amount of application in composites, for use in consumer oriented industries, is E glass fiber. Although the mechanical properties of E glass fibers are inferior to those of graphite and Kevlar®* fibers, as indicated in Table 5-4, the combination of properties and economic considerations always amounts to a distinct advantage for the use of these low cost fibers. Certainly, graphite fibers may find very limited usage in special applications such as drive shafts, which will be discussed later, but such usage will only be tolerated on an absolutely necessary basis. Kevlar fibers are unlikely to be utilized in structural applications. Glass fibers will be used in both continuous and chopped fiber form and, in most structural applications, as combinations of the two types.

The resin matrices for the composites will be the polymers which result in optimized composite cost/performance/processability. In the thermoset class of resins, polyesters are unquestionably the favorites, followed by vinyl ester resins and, for special applications, epoxy resins. The popularity of polyester resins arises from the favorable cost/processability advantages. The rationale for thermoplastic matrices is much less clear, since developments in this area lag far behind composites utilizing thermoset resins. It appears likely that existing types utilizing, for example, polypropylene (e.g., Azdel)† will be complemented by thermoplastic materials exhibiting somewhat better translation of composite properties and higher temperature presistance, for example polyethyleneterephthalate (PET).

Current applications of semistructural composites of the sheet molding com-

Table 5-4. Comparison of Fiber Properties.

Fiber	Strength (10^3 psi)	Modulus (10^6 psi)	Elongation (%)	Cost ($/lb)
E glass	350	10	3.4	0.55
S glass	500	13	4.0	2.30
Kevlar	400	18	2.2	9.00
Graphite, HS	400	32	1.2	32.00

---

*Registered trademark of E. L. du Pont de Nemours & Company.
†Registered trademark of PPG Industries.

pound (chopped glass) type usually involve an exterior surface. For example, the skin panels on a Corvette. By far the largest applications of sheet molding compound (SMC) is in grille opening panels, an illustration of which is given in Figure 5-2. Virtually all North American cars have SMC grille opening panels. These materials are compression molded and are generally designed as having orthotropic properties, although obviously after compression molding some degree of anisotropy may occur in the final product. SMC finds such widespread usage in grille opening panels not because the material itself is inexpensive (it is clearly more expensive per pound than mild steel), but because of the decrease in assembly costs resulting from the one-piece molding in SMC replacing several individual pieces of stamped steel which have to be joined together in a subsequent assembly operation. The total system cost, therefore, is reduced. However, one critical requirement of all exterior parts on a vehicle is the quality of the surface finish. Thus, although SMC exterior panels may well result in a weight save, the inability of these materials to produce the so-called Class A surface finish has limited SMC usage in such applications. The achievement of a Class A surface finish on an SMC part has been the subject of research and development within the SMC industry for the past decade. While great strides have been made, it is generally recognized that the best surface finish to be achieved on these materials is still somewhat inferior to the surface which can be obtained on a stamped steel part. The extension of the use of chppped glass SMC type materials to truly structural applications, i.e. elements of the structure which bear the major durability or crash loads within the vehicle, is inappropriate since the properties of these materials are inadequate to meet the stringent requirements of such structures. Table 5-5 lists typical properties for

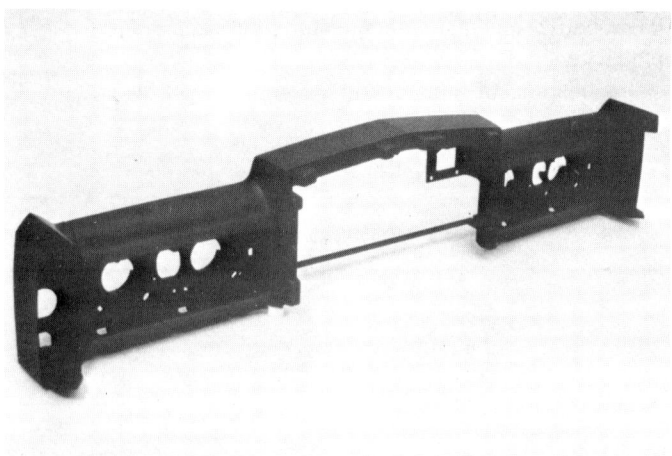

Figure 5-2. Typical SMC grille opening panel.

Table 5-5. **Typical Properties of SMC Materials.**

Material	Wt % Glass Fibers	Flexural Strength (10^3 psi)	Flexural Modulus (10^6 psi)
SMC-R30	30% chopped	30	1.8
SMC-R65	65% chopped	50	2.0
XMC	65% continuous	130	5.0

SMC type materials. Obviously, these materials must contain continuous glass fibers in addition to the chopped glass fibers in order to fulfill the structural requirements of the major load bearing structures of the vehicle.

At first sight, it might appear that the design criteria for the use of high performance composites in automotive structures would be less stringent than those currently utilized in aerospace type applications. In fact the reverse is true. The geometric complexity of components and structures in an automobile and the varied type and direction of the load inputs into the structure lead to a complexity of design, which is not equalled by any type of aerospace application. Rarely are automotive components symmetric or regular in shape and loading. This leads to a fundamental conflict with the governing rule for designing with composites. This rule, which can be summarized as "make sure the fibers take all the loads," would lead to inefficient designs in the complex loading environment of an automobile.

The degree of sophistication of the design criteria for the application of composites to the load bearing structures must clearly be of the level of that attained by the aerospace industry. In fact, these techniques must be expanded to allow for the complex stress patterns on typical automotive components. The cost restrictions imposed on the consumer markets mandate that the composites be utilized to maximum efficiency in order to be cost competitive and, in addition, that they be processable by mass production techniques. These restrictions impose a formidable task, but the challange is being met head on and promises to lead to innovations in automobile construction.

There appear to be two distinctly different types of design procedures for composites, that will be necessary to optimize their usage in the automobile industry. First, there are a select few components which can directly utilize the type of design procedure currently employed in the aerospace industry. Examples of such components are leaf springs and drive shafts. Typical examples are shown in Figures 5-3 and 5-4. If we consider, for example, a typical leaf spring manufactured in composite materials, it would consist of unidirectional glass fibers along the length of the spring. Thus, utilization of standard lamination theory and typical glass fiber/epoxy properties would lead to the design of a fail-safe spring in which both the spring rate and the rated load are satisfied. However, design calculations must also account for a twisting moment which

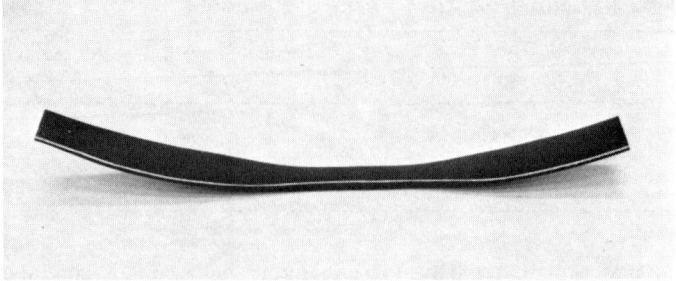

Figure 5-3. Composite leaf spring—"bow tie" design.

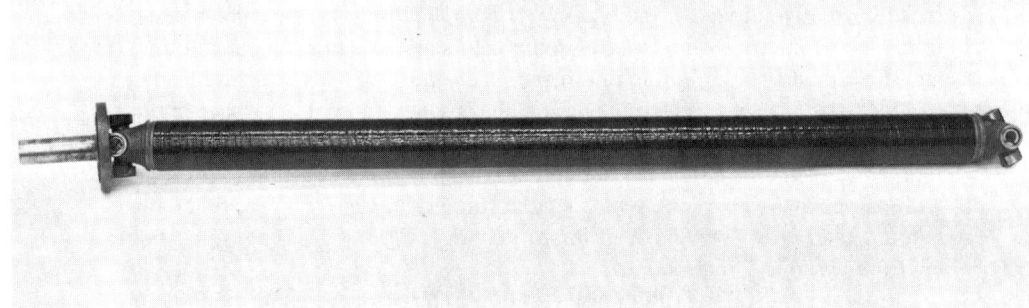

Figure 5-4. Composite drive shaft.

results in a shear stress in the spring, as well as for impact loads and any types of adverse treatment that may be encountered in the field. The design is also, of course, very sensitive to the manufacturing process since it is possible to fabricate leaf springs by a variety of techniques. For example, in Figure 5-5, three different types of leaf springs are shown, all of which will satisfy the requirements imposed. However, the efficiency of the design will vary with the type of spring chosen, as will the loading, the resulting bend, interlaminar and intralaminar stresses. Glass fiber reinforced epoxy leaf springs are a marketplace reality and have proven to be reliable in several applications over the past few years.

In a similar vein to leaf springs, drive shafts also utilize essentially direct derivatives of aerospace-type laminate design programs. In general, however, composite drive shafts (Figure 5-4) will only find application in special circumstances. The reason is an economic one resulting from the fact that the longitudinal stiffness of a drive shaft necessitates the application of graphite fibers in the longitudinal direction. Glass fibers can be uilized at $\pm 45°$ to take the

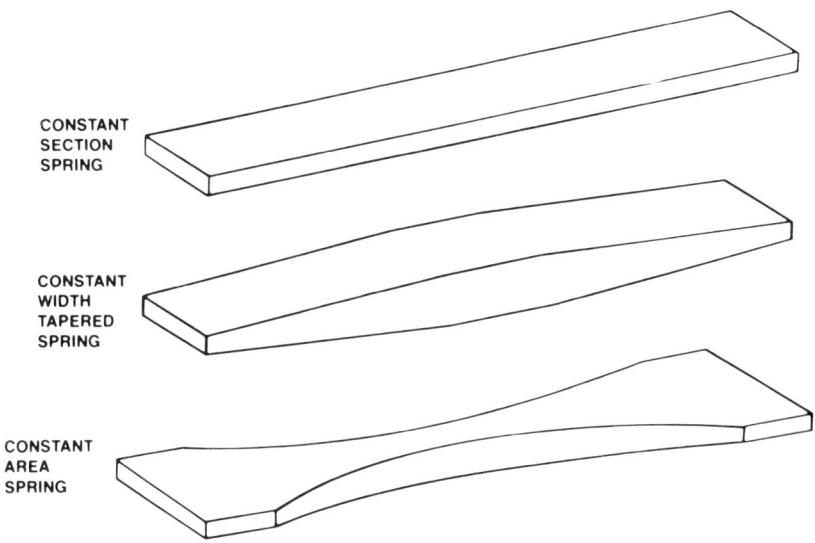

Figure 5-5. Different composite leaf spring configurations.

torque loads. Utilization of graphite fibers for bending stiffness increases the cost of such drive shafts to the extent that they are only competititve in a replacement for a two-piece steel drive shaft.

The second major area of design requirements involves large complex structures, for example, body structures in a vehicle. Depending on the manufacturing process, the utilization of composites in such structures will involve a combination of both continuous glass and chopped glass fibers. The uniformity or localization of the continous glass will be a sensitive function of the process by which the structures are manufactured. Consequently, the design program is required to be able to handle a mix of these two types of materials such that all the major loads are carried by the continuous glass fibers, while some degree of orthotropic behavior is manifest by the chopped glass system. Optimization procedures are required to produce the best combination of cost/properties/manufacturing. The design allowable are also a sensitive function of the manufacturing process and must be determined for all types of fabrication techniques in order to allow safe designs. Data base development and management, therefore, become primary requirements for successful applications of these types of composites.

Perhaps the biggest single requirement in the area of design criteria and methodology is the development of a simplified process of design, which is amenable to use by engineers who are not experts in materials. Virtually all aerospace design is done by hand, and the computer programs are confined to analyses to ascertain the safety of the structures developed by the composite experts. In consumer-type applications, the design methodology must preferably be a *design and analysis* procedure in which the computer program will do the design itself, perform the analysis, and iterate to develop the optimum design with the restrictionof the manufacturing constraints placed upon it. Again, this major undertaking is being pursued by the automobile companies in order to facilitate the use of high performance composites without a complete retraining of a large engineering community. This major deviation from aerospace-type techniques is not a trivial task. It is, however, a critical one to remove the aura of complexity with surrounds the design and analysis of composites, and these materials must be capable of being considered on the same basis as high strength steels or aluminum.

Quantitative design criteria must be developed in at least one new area to ensure efficient usage of composites. Integration of small components into one large composite structure is a key feature of the economic use of these materials. Clearly, the contribution of the joints eliminated by this integration to the basic properties of the composite material must be quantitatively evaluated to apply these structures efficiently.

FUNCTIONAL CONSIDERATIONS

There is a wide variety of functional considerations involved in structural applications in automobiles. Probably the critical areas from a structural viewpoint are durability, energy absorption, and NVH (noise, vibration, and harshness characteristics of the vehicle, which are perceived and rated by the occupants). These three requirements will dominate the design requirements for all structural areas of the vehicle. Inevitably, most areas of the structure will have to satisfy all three criteria to some degree, but one requirement will be the primary one. For example, many suspension components are durability (fatigue) critical and thus are designed to satisfy this requirement; however, the component may also play a role in crash response and NVH characteristics. (As mentioned earlier, it should always be remembered that surface finish is a critical consideration if vehicle exterior surface is involved.)

An evaluation of the anticipated increased damping of composite structures relative to vehicle NVH must be quantified. The vibrational characteristics of a vehicle, which are dictated by stiffness and damping, contribute significantly to the passengers' perception of comfort. It is anticipated, and at least qualitatively confirmed from leaf spring experience, that widespread use of composites should diminish vibrational response due to increased damping. However, the extent to which this is perceptible to the occupants has yet to be determined.

Extensive testing programs are underway to determine the mechanical properties, in particular the fatigue (durability) characteristics, of a wide variety of

Table 5-6. Typical Bend and ILS Fatigue Data.

Material	Bend Fatigue Strength / Bend Strength	ILS Fatigue Strength / ILS Strength
SMC-R30	0.35	
SMC-R65	0.3	
XMC	0.3	0.6
0° Glass/Epoxy	0.35	0.68
0° Graphite/Epoxy	0.8	0.72
0°/90° Graphite/Epoxy	0.75	0.58

glass fiber reinforced composites, which have potential application in automobile structures. Although this task is awesome in magnitude because of the added complexity of environmental and temperature sensitivity, the determination of such properties is relatively straight-forward in that this type of testing in other materials has been going on for decades. The development of adequate fatigue prediction procedures (e.g., for cumulative damage), which are necessary for efficient application of composites, presents a challenging and difficult task.

Typical fatigue data for a variety of composites are given in Table 5-6. In addition to the conventional tensile (bend) fatigue operative in all materials, composites have the additional complexity of exhibiting interlaminar shear (ILS), fatigue which is unique to fiber reinforced materials. Data for ILS fatigue are also included in Table 5-6. Note that all the composites appear to have a high resistance to ILS cyclic stresses.

Another area of data accumulation represents a somewhat different task. The requirement that vehicle structures absorb energy in high speed impact to protect occupants from injury, involves extensive knowledge of the way composite structures deform and collapse under high speed conditions. Obviously, since composites are essentially linearly elastic in nature, these materials will not absorb energy in the same manner (namely, plastic deformation) in which steels and other metals achieve this. Rather, composites will absorb energy by controlled disintegration or fracture. Thus, the plastic deformation energy mechanism of metals is being replaced in composites by a fracture energy asorption mechanism. Figure 5-6 illustrates the difference between the two processes. It has been clearly demonstrated, particularly by Thornton and coworkers [1, 2], that composites can absorb significantly more energy per unit weight than steels in axial collapse. Figure 5-7 shows a comparison of energy absorption in a steel tube and energy absorption in a graphite fiber reinforced composite. Thus, under carefully controlled laboratory conditions, composites can be more weight effective than steels for energy absorption. However, in the complex geometric

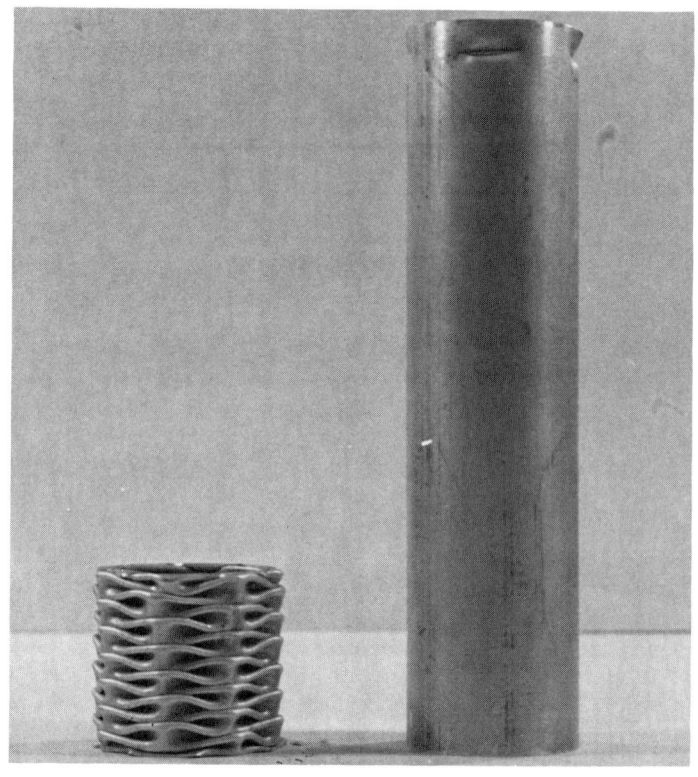

(a)

(b)

Figure 5-6. Typical axial collapse mechanisms for (a) metal and (b) composite tubes.

AUTOMOTIVE APPLICATIONS OF COMPOSITES 185

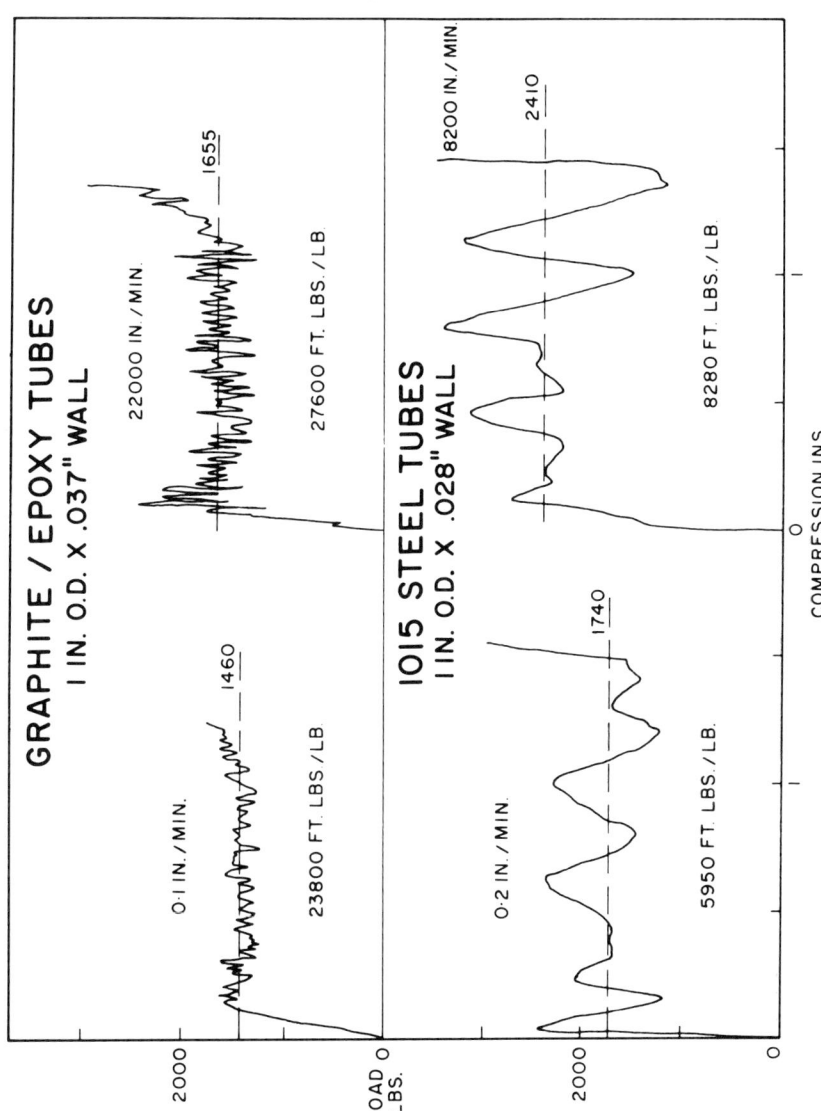

Figure 5-7. Comparison of energy absorbing capability of metal and composite tubes.

structure of a vehicle, care must be taken to ensure that instability criteria are not violated, otherwise low energy absorption can arise. Clearly, this is also true of steel structures, and in fact, indications from laboratory data suggest that similar geometries in composites tend to be more stable than in steel. This whole area of energy absorption in complex geometric structures is only currently being evaluated, although there is no reason to suspect that the weight efficiency of composites exhibited in simple structures will not be carried over to these more complex structures.

METHODS OF MANUFACTURE

Injection molding of short glass fiber reinforced thermoplastic parts presents, perhaps, the most rapid polymer processing technology available for nonstructural parts. Molding cycle times are on the order of fractions of minutes, and multiples of the part are made concurrently. Basically, a fluid polymer carrying with it the chopped glass fibers is injected into a closed, cooled mold. As soon as the polymer has cooled sufficiently to become rigid, the mold is opened, the parts are ejected, and the cycle repeats itself. The process is rapid and inexpensive enough to be used for many small parts, in particular tending to replace parts originally made by zinc die casting. Again, it should be emphasized that these components made by injection molding tend to be decorative or semistructural at best. In the last few years, the advent of reaction injection molding (RIM) and reinforced reaction injection molding (RRIM) has led to the use of this process for exterior panels, particularly for vertical surfaces. The development of the so-called friendly fender is based on the RRIM process. As currently envisaged, injection molding does not lend itself to the fabrication of parts which undergo major load bearing environments.

A second major manufacturing process, which is widespead throughout the automobile industry, is compression molding of sheet molding compound (SMC). A schematic of the SMC process, indicating both the fabrication of the SMC material itself and the subsequent compression molding into a component, is shown in Figure 5-8. As mentioned earlier, the SMC technology is widely used in the automobile industry for the fabrication of grille opening panels on all vehicles and for some exterior body components on selected vehicles, for example, tailgates (Figure 5-9) and heavy truck cabs (Figure 5-10). The fabrication process consists of taking sheets of the leathery textured SMC material (which is made up of chopped glass fibers in uncured resin), cutting predetermined shapes of the SMC, dropping the shapes into the heated mold, and closing the mold under pressures of approximately 1000 psi. Depending on the exact formulation of the SMC, a three-minute cycle at a temperature of approximately 270°F (typically) is characteristic of the process.

The same basic operation described for the manufacture of components out

SHEET MOLDING

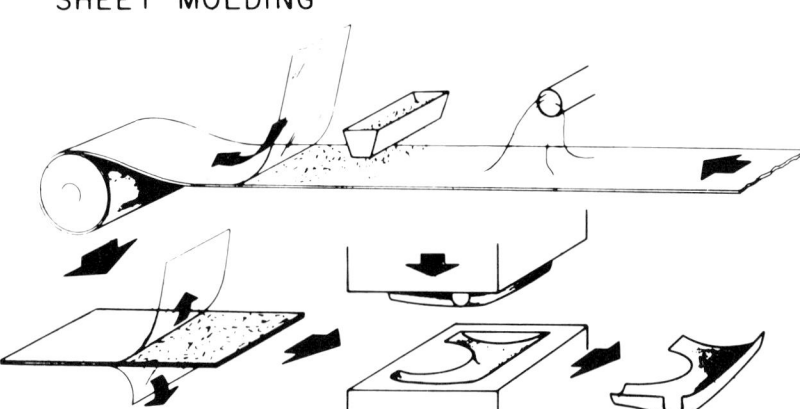

Figure 5-8. Schematic of SMC manufacture and component fabrication.

of SMC can be used for combinations of chopped glass and continuous glass. There are two variations by which this might be achieved. By feeding in both continous glass strands and chopped glass from a chopping operation, the same machine which produces the SMC material (shown in Figure 5-8) can be used to produce material containing alternate layers, or mixtures of layers, of chopped glass and continuous glass. The resultant material can be tailored to produce whatever combination of properties is desirable between all chopped glass and

Figure 5-9. SMC tailgate for Ford Bronco II.

Figure 5-10. SMC heavy truck cab.

all continuous glass. The material is then cut to shape in a manner exactly analogous to that already described, dropped in a heated die and compression molded. The second variation on this technique is to make SMC containing all chopped glass and then, on a similar machine, make an analogous product with 100% continuous glass (XMC). During the fabrication process, mixtures of the SMC and XMC are laid into the mold so that the continuous fiber material, XMC, is in the required places to carry the maximum load. In both these operations, it must be understood that whereas in the compression molding of chopped glass SMC the material flows to fill the mold, in material containing continuous fiber the flow is restricted in the longitudinal direction of the continuous fibers, although some degree of flow is exhibited perpendicular to the fibers. Thus, the complexity of shape and the amount of flow necessary to completely fill the mold are sensitively dependent on the location and amount of the continuous glass fibers. Obviously, for simple shapes, this does not present a problem. However, the presence of any significant geometry in the part requires careful development of material patterns prior to charging the mold to ensure an integrally molded part having the required properties. Figure 5-11 shows a prototype rear floor plan of an Escort molded in one piece, where the channel sections contain continuous fibers because of the required strength and stiffness, and the rest of the load floor consists of chopped glass SMC material. Development of the charge pattern led to good interpenetration between the XMC and SMC material, resulting in excellent load transfer and adequate material flow to fill the mold. This particular application, being a prototype, was a direct geometric duplicate of the steel floor pan which consisted of ten individual steel

Figure 5-11. Prototype rear floor pan molded in a combination of SMC-R65 and XMC.

parts joined by spot welding. Obviously, it would be beneficial to completely redesign such a structure to facilitate the fabrication and allow for the different degrees of flowability of the continuous XMC and chopped glass SMC.

Compression molding SMC technology is a powerful fabrication process which is economically viable today. However, one of the disadvantages of this process is that the degree of flow required to optimize the properties (remove air) results in a spread in mechanical properties of the material due to nonuniform flow throughout the mold. For safety, the design allowables must be based on the minimum properties, since they represent the weak link in the structure. In a molding such as that shown in Figure 5-11, variations in mechanical properties by a factor close to 2 within the chopped glass SMC area of the components can typically be found, and some areas are probably overdesigned by this same factor. Thus, while compression molding may be economically efficient, it may result in some structural inefficiency and weight penalty. Great efforts are currently being made to minimize the spread in properties in a typical molded part by flow modeling of such structures to optimize charge pattern shape and distribution, so that fiber deficient regions will be minimized and knit-line locales will be situated in low stress areas. Obviously, the direct results of such optimization should result in both cost and weight minimization.

It is clear from the comments on SMC technology that fiber management is the critical factor in composite processing. If we take this argument to a logical conclusion, prepositioning of the fibers in a mold, prior to the introduction of the resin, would therefore represent the optimal processing technology for complex structures for automobiles. Such processing (preform molding) would result in maximum reliability, minimum weight, and tailored properties. Resin transfer molding represents just such a procedure. The technique involves placing glass fibers in the appropriate mix and orientation inside of mold, closing the mold, and then injecting a low viscosity resin to fill the mold without disturbing the fiber placement. Until recently, one of the major drawbacks has been the inability to fill the mold rapidly enough without perturbing the placement of the glass fibers. However, low viscosity resin technology promises to overcome this drawback and allow relatively rapid processing time while guaranteeing fiber positioning. Perhaps the only remaining obstacle to large scale usage of such a process is the automation of fiber placement in the mold. The commonly used procedure of hand-laying the mixture of fibers throughout the mold is obviously labor-intensive, and resolution of this expensive part of the fabrication process would result in an ideal procedure for making high performance composite structures for consumer market.

One other developing fabrication process deserves mention in terms of future potential in the automotive industry, namely, thermoplastic stamping. The process consists of heating reinforced thermoplastic sheets, stamping the materials in matched metal dies, and allowing the material to cool in the dies before extracting the part. The advantages include rapid cycle time and infinite material shelf life (by contrast, thermoset based materials like SMC have a restricted shelf life). Fiber reinforced materials of moderate stiffness and strength such as glass fiber reinforced polypropylene (Azdel) are already used for structural parts (e.g., seat backs, battery trays), and further developments in these materials will lead to their increasing use in vehicle structures.

One key aspect of manufacturing, although not a material fabrication process, must be taken into consideration. Vehicle assembly in the atuomobile industry is largely one of joining processes. The spot welding of steel components into an integrated structure is a well-developed technology involving one of the major robotic development areas. If large segments of a vehicle structure consist of composites, rapid and reliable techniques for joining composites to other composites, and composites to steel components, must be in hand as part of the manufacturing and assembly process. Inherent in this assembly process is the criterion that speed is of the essence. There are two generally accepted methods for joining fiber composites, both to themselves and to dissimilar materials. These techniques are adhesive bonding and bolting. Adhesive technology is fairly well developed by usually requires significant cure time. Adhesive joints require special consideration during the design of the structure since, while usually of adequate strength in shear loading, such joints are sensitive to peel stresses or cleavage loads. Care must be taken to eliminate or minimize these stresses. To facilitate the curing time, a combination of both mechanical joints and adhesive bonding will tend to be used.

A rapid technique for inspection of composite structures and joints is another requirement, particularly in the early years of application, to ensure quality control. Nondestructive test techniques (NDT) under development include ultrasonic C-scan, radiography, acoustic emission, and holography. Although all these techniques can be utilized on a laboratory scale, ultrasonic procedures are probably the best for adaptation to a production environment.

POTENTIAL OF COMPOSITES

The Ford graphite fiber reinforced plastic (GrFRP) concept vehicle built in 1979 Figure 5-12 demonstrated perhaps the ultimate potential of high performance composites in automobiles. The project was intended to demonstrate concept feasibility and to identify critical issues related to production feasibility for future vehicles. Manufacturing and cost feasibility were not program objectives.

The completed vehicle weighed 2504 pounds, a savings of 1246 pounds relative to the production steel car. A summary of the weight savings—typically, on the order of 60%—is given in Table 5-7. The experience with this concept vehicle emphasized the two key issues which will determine future use of composites—materials cost and manufacturing feasibility. The materials cost issue will dictate glass fiber usage, barring an unanticipated revolution in graphite fiber prices. Low cost manufacturing processes are technologically feasible.

Figure 5-12. 1979 Ford LTD constucted from GrFRP.

Table 5-7. Major Weight Savings in GrFRP Ford.

Component	Weight (lb) Steel	Weight (lb) GrFRP
Body-in-white	423	160
Front end	95	30
Frame	283	206
Wheels	92	49
Hood	49	17
Decklid	43	14
Doors	141	56
Bumpers	123	44
Drive shaft	21	15

The GrFRP vehicle project enhanced the belief that composites can, and will, play an increasingly important role in the automotive industry. Virtually all areas of the vehicle are being extensively studied for potential composite usage. Part integration, simplified assembly, and reduced investment costs are becoming increasingly important considerations in addition to weight savings, and the total systems benefit accruing from the use of high performance composites—rather than any single advantage—will be the key to their usage. Close coordination and interaction of design, manufacturing methods, and materials engineering in an integrated systems approach will be vital. The commitment of the automotive industry to develop the necessary technological innovations in unquestionable. The next decade will witness dramatic changes in the usage of composites.

References

1. Thornton, P. H., *J. Comp. Mat.*, **13** 247–262 (1979).
2. Thornton, P. H., and P. J. Edwards, *J. Comp. Mat.*, **16,** 521–545 (1982).

6
COMMERCIAL AIRCRAFT APPLICATIONS

John T. Quinlivan
J. Corey McMillan
Materials Technology Research and Development
Boeing Commerical Airplane Company
Seattle, Washington

In the early 1980s, a new generation of large commercial jet aircraft emerged on the world scene, with the introduction of the Boeing 767 and 757, and the Airbus Industries A310. These were the first all-new large jet designs in nearly a decade. As would be expected, these new aircraft employed the latest state of the art in a number of high-technology areas, including materials. The new materials incorporated in these aircraft included the first widespread application of advanced composites in commercial aircraft production.

Within the comercial aircraft industry, advanced composites are generally defined as the fiber-reinforced composite systems that followed fiberglass. In actual airframe usage today, these new systems are generally limited to carbon-epoxy, aramid(Kevlar™)*-epoxy, and hybrid systems of carbon and aramid or sometimes glass fibers.

Although this chapter focuses on the new advanced composites, it also covers the early usage of fiberglass, which was the precursor of current technology. In addition to past and present usage, projections for future usage are addressed. Forecast requirements for advanced composites project dramatic increases, with the aerospace market playing a dominant role.

---

*Registered trademark of E. I. du Pont de Nemours & Company.

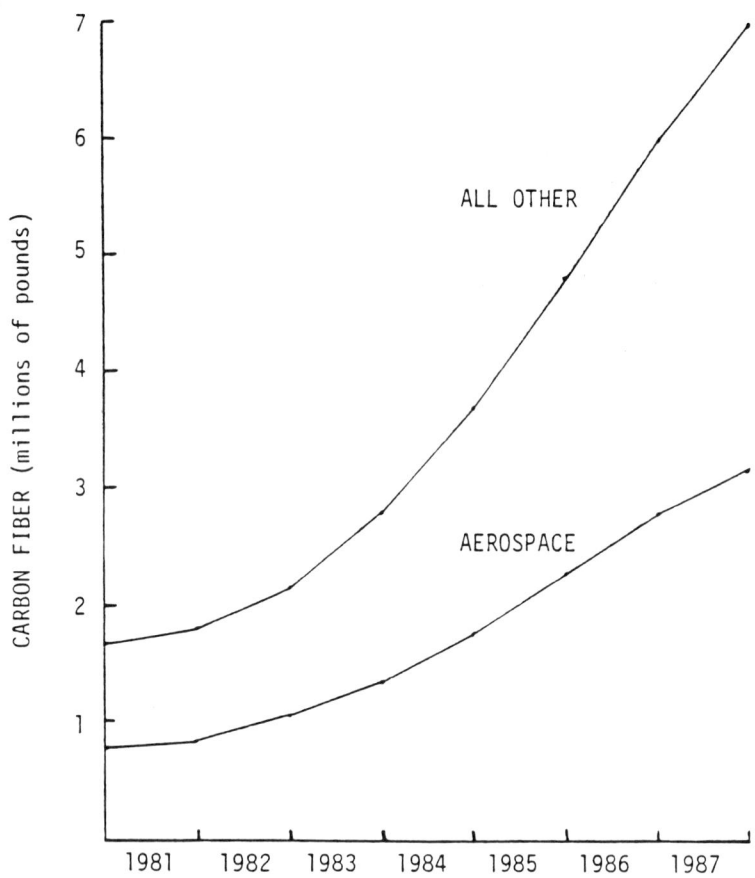

Figure 6-1. Carbon fiber requirements as estimated in October 1982 [1].

These increases are highlighted in Figure 6-1, which shows the past demand and estimated future requirements for carbon fiber. These data are drawn from a market survey report [1] that presents detailed information. The general trends project a 12 to 13% growth per year in total as well as aerospace requirements, with aerospace nearly half the total. The commercial aircraft use parallels these predictions. Within the industry, however, there is some disagreement, with optimistic forecasters predicting annual growth rates as high as 30 to 40% per year. The clear consensus on composite usage is an increase: the questions are the degree and the timing.

EARLY COMPOSITE APPLICATIONS

Fiber-reinforced composite materials have been incorporated in aircraft since the first days of flight. The earliest composites were doped linen fabric coverings over spruce lumber. Wood is a composite of cellulose fibers in a lignin binder. Doped linen fabric was employed in the construction of the aerodynamic surface components of fighters and some large bomber/transport aircraft until well into World War II. Portions of the F4U-5 and -7 Corsair remained fabric covered until after the war, when the components were changed to metal.

The first major use of composite structures was brought about by the advent of radar and the need for a radome material that was transparent to electromagnetic waves. Apparently, the first operational radome was made of plexiglass, frosted to prevent visibility of the antenna, and flown on a Boeing B-17 [2]. At about the same time, radomes made in Britain consisted of dry fiberglass laid up on a male mold that then was inserted into a female mold; polyester resin was poured between them to create the composite [3]. Another early application of reinforced plastic structures, circa 1945, was the development of glass-reinforced polyester resins for air-distribution ducting. These complexly shaped parts are easily manufactured on breakaway plaster tooling and are significantly cheaper than their metal counterparts.

The use of reinforced plastics on aircraft continued to grow as cost and design advantages were realized. By the mid-1950s, the wet impregnation method, which often required multiple-stage processing, was replaced by the use of resin-preimpregnated fabrics, or prepregs, thereby further reducing the cost of manual processing. These fabrics had controlled resin content, which reduced the weight of reinforced plastics as well. In the 1960s, single-stage processing of honeycomb sandwich structure was developed, which further reduced the costs and significantly increased the use of reinforced plastic materials.

In single-stage processing, resin-preimpregnated fabric (prepreg) is laid up together with all other parts of the assembly and the entire layup is cured as a single unit. This contrasts with wet lay-up systems, where dry reinforcing fabric and resin are brought together and cured to produce simple laminate structures. Assemblies then are produced by bonding or mechanically fastening the various simple laminates, similar to metal fabrication techiques.

Sheet metal and composite sandwich concepts for the construction of fairing sections are contrasted in Figures 6-2 and 6-3. Figure 6-2 shows a typical sheet metal skin and rib structure assembled with mechanical fasteners. A similar fairing, fabricated with glass-reinforced plastic honeycomb sandwich, is shown in Figure 6-3. Woven glass fabric reinforcements, preimpregnated with a 250°F-curing, controlled-flow epoxy resin, are used for the skins and doublers. The honeycomb is glass fabric or Nomex™* paper impregnated with a heat-resistant phenolic resin. The skin is bonded to the core by the epoxy resin in the prepreg skins. The resin is formulated to flow at a temperature just prior to gelation,

---

*Registered trademark of E. I. du Pont de Nemours & Company, Inc.

196 ADVANCED THERMOSET COMPOSITES

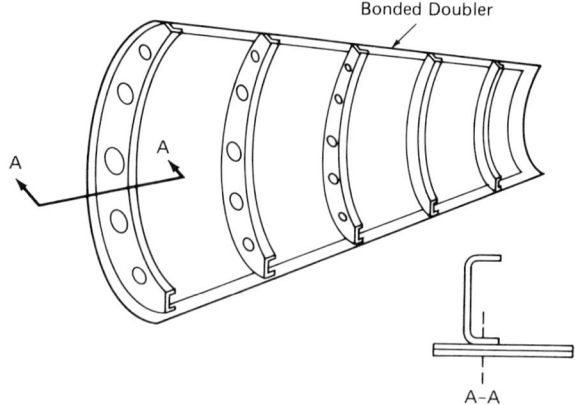

Figure 6-2. Fairing section—sheet metal construction. Tools required include one stretch form block, five hammer form dies for stiffeners, and one drill jig and assembly tool.

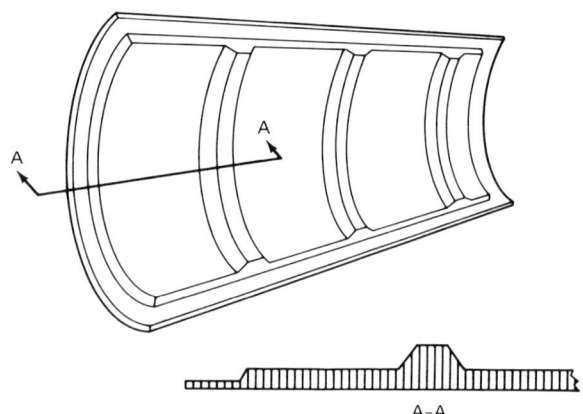

Figure 6-3. Fairing section—glass-reinforced plastic sandwich construction. Tools required include one female mold for outer contour and one template for locating honeycomb core and beams.

forming a fillet between the honeycomb cell wall and the skin. This is why the resin is called a "controlled-flow" resin.

The manufacture or lay-up operation is simple. The required number of prepreg layers are placed in the mold for the outer skin. Doublers are laid up against these plies. The prepared honeycomb core is positioned against the outer skin prepreg, followed by additional doublers and final inner skin plies of prepreg. The entire lay-up is enclosed in a vacuum bag and cured in an autoclave. Final trimming and finishing of the cured detail, after removal from the tool, are the only steps remaining before installation.

A single-step sandwich panel fabricated with prepreg skins and honeycomb core offers the advantages of lower manufacturing cost, improved strength, and reduced weight, as compared to multiple-stage-processed composites or metal parts. In addition, it is possible to produce panels with multiple contours, which is very expensive with the precured plastic skins. For the fairing sandwich shown in Figure 6-3, the use of the single-stage lay-up reduces labor hours by as much as 50%; in addition, the finished part is up to 34% lighter than the metal assembly it replaces. Many variants of this process sequence have been developed, including the use of adhesives for bonding to honeycomb cores or to metal details inserted within the lay-up.

So successful was the single-stage operation for fairing applications that the growth of fiberglass fabric usage on Boeing commercial aircraft reached a maximum of 10,000 square feet, or 25% of the exposed surface area, on the 747. Wing-fuselage fairings, rudder and elevator sufaces, wing leading and trailing-edge panels, and the radome make up 1200 pounds of cured honeycomb sandwich on each aircraft. The production of over a million square yards of fiberglass fabric-reinforced plastics for Boeing commercial aircraft served as a significant data base for the later development of similar carbon and aramid components [4].

Concurrently, glass fabric was finding its way into similar secondary structural applications in a wide range of other aircraft. Large wing-fuselage fairings were fabricated for the Lockheed L-1011 [5] and the original Airbus [6].

In addition, a large amount of glass cloth is used in the interiors of commercial aircraft. Nonstructural parts, such as the overhead luggage compartments, ceiling panels, galley partitions, bulkheads, and toilet enclosures, are routinely made of glass fabric laminates.

Two examples of interior fiberglass honeycomb components are shown in Figures 6-4 and 6-5. The 747 sidewall and parition panels are a light-weight sandwich construction incorporating a Nomex core overlaid with epoxy resin glass cloth. Because these panels have decorative or aesthetic requirements, they incorporate embossing resin and Tedlar™* layers. The Tedlar surface also is easy to clean—another panel requirement.

The glass floor panel is another example of a lightweight, honeycomb core-stabilized panel. These panels are laid up in large sections, with the details of the glass reinforcement defined by the specific load requirements. These load requirements vary with location: e.g., under-seat panels require less load-car-

---

*Registered trademark of E.I. du Pont de Nemours & Company, Inc.

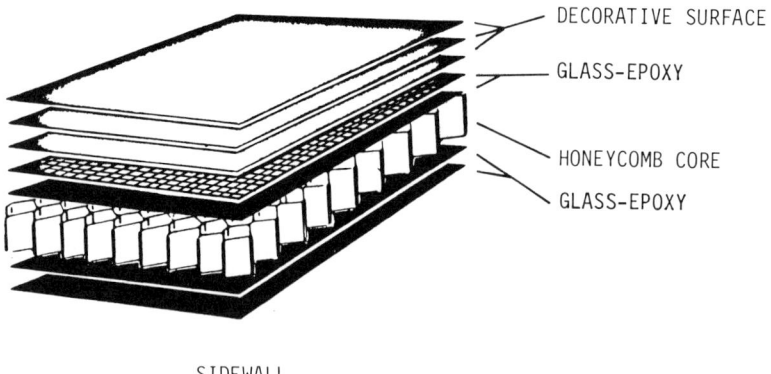

SIDEWALL

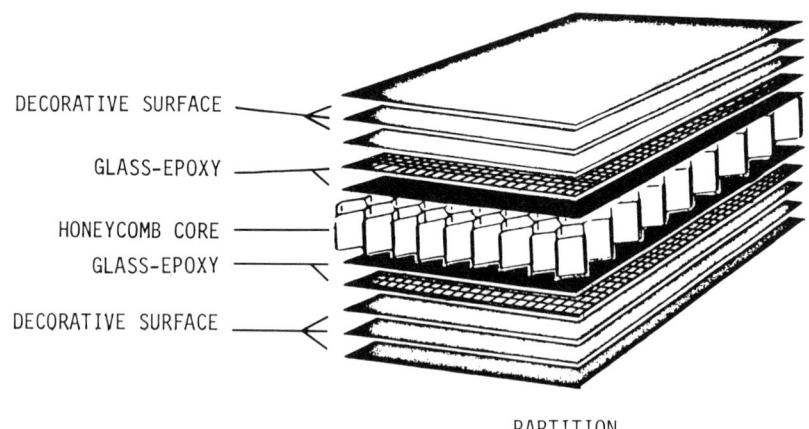

PARTITION

Figure 6-4. Interior sandwich panels.

rying capability than do aisle panels. The panels are cut to size, fastener attachment inserts are installed, and the edges are potted prior to installation.

In each of these components, the major driving force behind the use of composites was reduced weight. However, improved decorative schemes for the sidewell panels and improved durability for the floor panels also were important.

INTRODUCTION OF ADVANCED COMPOSITES

In the late 1950s and early 1960s, several new fibers with impressive structural properties were developed. Primary among these were boron, graphite and car-

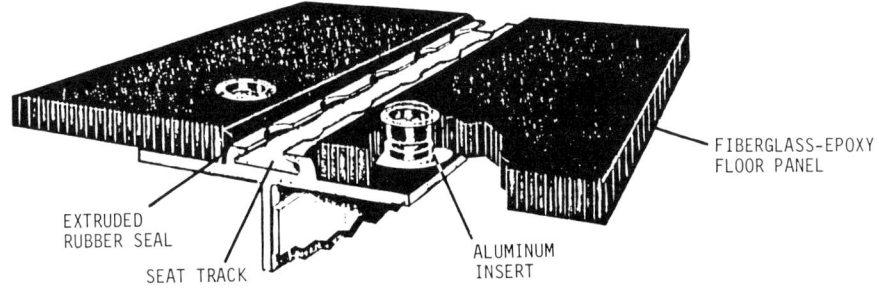

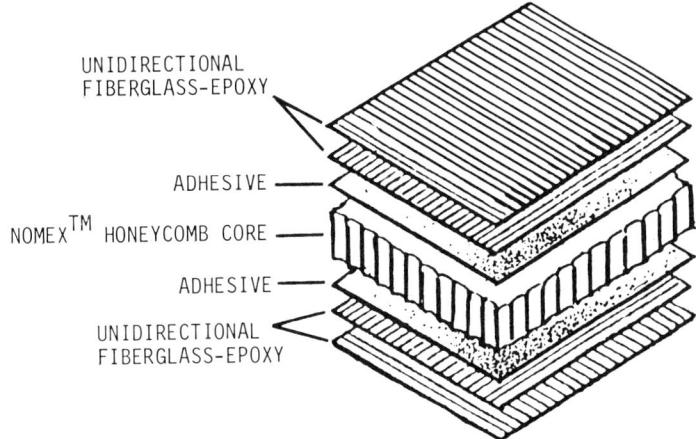

Figure 6-5. Fiberglass-epoxy floor panel.

bon, aramid, and S-glass. Composites of these materials possess very high specific strength and/or modulus, making them attractive candidates for weight reduction applications. By the late 1960s, testing and development had clearly identified carbon filaments as the fibers with the best overall balance of engineering properties, manufacturability, and cost. Because carbon fiber cost is still relatively high, however, various material combinations—often including glass or aramid fibers—are used to achieve maximum cost effectiveness.

Carbon fibers of the type discussed here often are referred to as "graphite" fibers in the trade and in the open literature. Graphite, by its technical definition, implies a three-dimensional crystallographic order, which is not present in the widely used, commercially available fibers. Hence, the term "carbon fiber" is technically correct.

Carbon fibers can be long and continuous, or short and fragmented. In general, short fibers cost the least and fabrication costs are lowest, but properties of the resulting composite also are lower than those obtainable with longer or continuous fibers. Where the ultimate in performance or weight reduction is required, continuous carbon fibers are usually the reinforcement of choice. Continuous fibers are available in a variety of forms, including yarns or tows containing from 100 to 160,000 individual filaments; woven rovings; braids; unidirectional impregnated tapes; preplied layers of tape with individual layers, or plies, at selected fiber orientations; and fabrics of many weights and weaves.

The individual carbon filaments are usually 5 to 8 microns in diameter. Tensile strength can exceed 600,000 psi; modulus, 100 million psi. Commercial fibers are available in three modulus ranges, which satisfy most design needs: 30 to 35 million psi, 50 to 55 million psi, and 70 to 75 million psi. Generally, the higher the modulus, the lower the strength (and, hence, strain). For example a strength of 400,000 psi is achieved at a modulus of 32 million psi, while fibers of 50 million psi modulus have strengths of approximately 300,000 psi.

Today's manufacturing technology can produce efficiencies approaching 100% in property translation. The degree to which the strength and modulus of a composite approach 100% of the corresponding filament properties (multiplied by the filament's volumetric fraction) is governed by the uniformity of filament properties, filament alignment within the composite, and integrity of the fiber/matrix interface.

The choice of fiber form often depends on the fabrication process selected. Filament-winding processes usually dictate the use of low-filament-count yarns to minimize the catenary effects (and resultant looseness of windings) common with high-filament-count yarns. Autoclave, vacuum bag, and compression-molding processes used for relatively flat or simple-curvature parts can employ unidirectionally oriented tapes, laid up in place or preplied. Fabrics commonly use low-filament-count yarns (1000 or 3000 end tows) to minimize weight and ply thickness.

Although woven fabrics are more expensive than unidirectional tapes, cost savings often are realized in the molding operation because labor requirements are reduced. Complex part shapes or processes requiring careful positioning of the reinforcement can benefit from the use of the woven forms of carbon fiber, which are easier to handle.

Some fabrics are essentially unidirectional. Most of all of the carbon fibers in these fabrics are oriented in one direction and are held in position by nonstructural tie yarns. Other commercial fabrics are of a plain or satin construction. Satin-weave fabrics, particularly the commonly used eight-harness satin, retain most of the fiber characteristics in the composite and can be easily draped over complex mold shapes. Plain-weave fabrics are less flexible and are suitable for flat or simply contoured parts, but at a slight sacrifice in fiber property translation.

The major drawbacks to using carbon fibers are low impact resistance and high cost. Both of these factors are improved by the use of hybrids (carbon fibers incorporated with lower-cost, higher-impact-strength reinforcements). Hybrids are available in several different forms. For filament winding or tape lay-up, the secondary reinforcement may be introduced as intermixed fibers within the same layer or as separate layers oriented to derive maximum benefits from each material. Commercial fabrics are available that contain a mix of fibers within the warp and/or fill, as well as the carbon in one direction and the secondary fiber in the other. The most common hybridization technique is to intermix plies of fabrics or tapes of different fibers. For example, carbon fiber often is used for stiffness and located only in the honeycomb bay areas of composite panel. Aramid or glass is used in all the edgeband areas, as well as in the honeycomb areas.

ADVANCED COMPOSITE SERVICE EVALUATIONS

A number of smaller commercial airplane parts and components have been designed and built of composites, largely as a result of NASA and Air Force interest in learning to use these materials. In July 1973, under the direction of the NASA Langely Research Center, a flight service program was initiated with the placement of 111 production 737 carbon-epoxy spoilers into service. Seven airlines using 28 aircraft in various parts of the world have participated in the flight service evaluation. Periodically, selected spoilers are removed from service for laboratory test. This program has contributed significantly to the understanding of the behavior of composite components in service. The spoilers have accumulated over 1.8 million flight hours and 2.6 million landings. A photograph of a typical spoiler installation is shown in Figure 6-6. Service data are summarized in Figure 6-7.

The spoiler design concept required fabricating spoilers that were interchangeable with the existing production parts. The design involved replacing the aluminum skins with carbon and the end-closure ribs with fiberglass. The rest of the parts—center hinge fitting, front spar sections, and honeycomb core—were the existing aluminum production parts.

A DC-10 floor beam and two supporting struts were among the first primary structural members selected for composite application. These components were chosen because the number of similar beams on an airplane was sufficient to produce significant weight savings and a high-volume manufacturing rate. The aft fuselage station 1541 beam was selected because it is located at the end of a fuselage barrel section, thus allowing the floor beam to be installed immediately before the aft fuselage is joined to the center fuselage section.

Installation of the 20 foot long beam is shown in Figure 6-8. The numerous holes in the beam are for routing electrical wiring; control cables; and hydraulic, pneumatic, and fuel lines through the aircraft. The composite floor

202 ADVANCED THERMOSET COMPOSITES

Figure 6-6. Boeing 737 flight spoiler.

Airline	Total Spoiler Hours	Total Spoiler Landings	Spoilers Currently in Service
Frontier	80,822	86,950	2
Aloha	174,791	444,994	0
New Zealand	238,388	324,620	11
Lufthansa	403,077	511,469	12
Piedmont	639,101	928,893	23
VASP	263,876	296,574	11
PSA	29,747	51,521	0
Total	1,829,802	2,645,021	59

Figure 6-7. Boeing 737 flight spoiler service summary (fall 1983).

COMMERCIAL AIRCRAFT APPLICATIONS 203

Figure 6-8. McDonnell Douglas DC-10 composite floor beam installation. Beam installation in airplane fuselage entered flight service in May 1979.

beam was designed to match the bending stiffness of adjacent floor beams and to sustain the critical design ultimate loads for both passenger and cargo configurations.

Lockheed-California Company has developed and tested a carbon-epoxy L-1011 inboard aileron under a NASA-Langley contract as part of the aircraft energy efficiency (ACEE) program. The aileron configuration is shown in Figure 6-9. The aileron incorporates carbon-syntactic (microballoon-filled) sandwich skins and carbon channel-section spars (made from tape) and ribs (made from fabric).

The McDonnell Douglas DC-10 upper aft rudder program, begun in 1974, resulted in the fabrication of 20 carbon-epoxy rudders for flight service. To date, 13 rudders have been installed on DC-10s. These rudders were fabricated as a unitized structure in a single cure cycle using an out of autoclave, expanding-rudder manufacturing method. The structural elements of the rudder are shown in Figure 6-10. The skins, ribs, and front and rear spars were consolidated into a monolithic structure in a single cure cycle without using adhesives

204 ADVANCED THERMOSET COMPOSITES

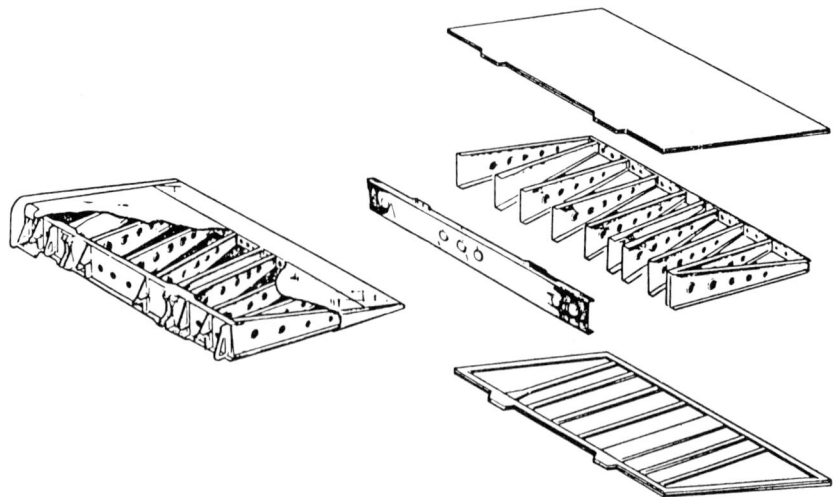

Figure 6-9. Lockheed L-1011 composite aileron configuration.

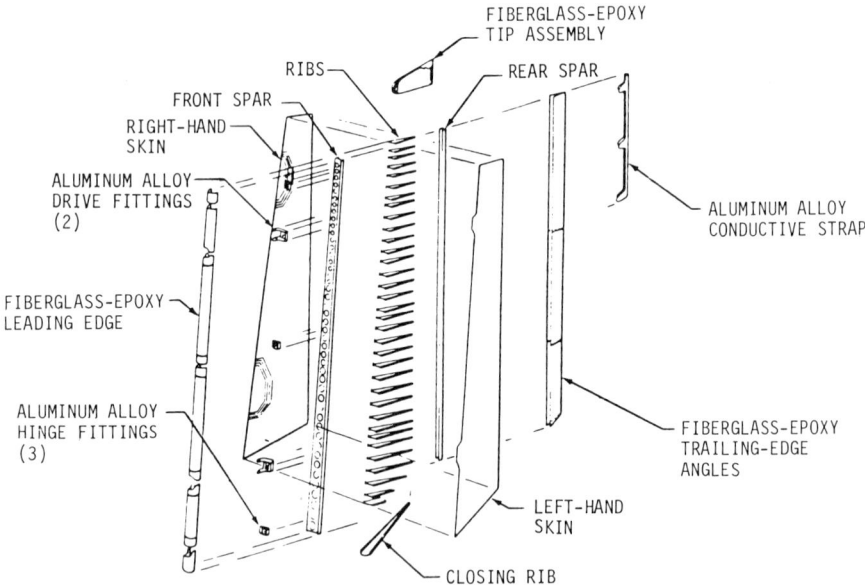

Figure 6-10. McDonnell Douglas DC-10 rudder structural arrangement. Skins, spars, and ribs are all cured together in a single operation.

COMMERCIAL AIRCRAFT APPLICATIONS 205

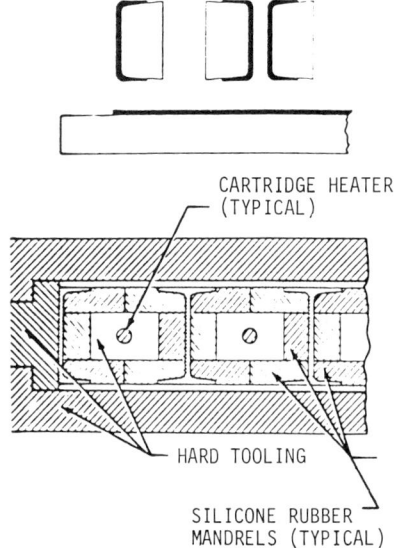

- LAY UP INDIVIDUAL PIECES ON ANCILLARY TOOLS
- PREFORM AND DENSIFY IN "B" STAGE WITH MODERATE HEAT AND PRESSURE
- TRIM TO SIZE
- STORE IN FREEZER
- ASSEMBLE PIECES IN CURING MOLD
- APPLY HEAT
- PRESSURE SUPPLIED BY EXPANSION OF SILICONE RUBBER WITHIN MOLD CAVITY
- NO BAGGING, BLEED-OFF, ADHESIVE, OR AUTOCLAVE REQUIRED

Figure 6-11. Thermal expansion molding process.

or secondary bonding. This thermal expansion molding proces is shown in Figure 6-11. Access to remove the inside tooling was gained through holes molded in the front spar. The process yielded a ready-to-trim rudder box of excellent quality.

Recently, under a NASA contract, the elevators on the 727 were redesigned using carbon-epoxy, tested, and certified for use on the commercial fleet. The composite parts, including the balance weights, are 23% lighter than the metal parts, a savings of 146 pounds per airplane. The 727 composite elevator is a complete redesign but retains component interchangeability. After extensive testing, five shipsets were placed into service starting in March 1980. A photograph of a 727 with composite elevators installed is shown in Figure 6-12.

During the 727 composite elevator design study stage, four preliminary composite elevator configurations were considered. Concept 1 used a minimum number of ribs and skin panels stabilized by Nomex honeycomb core. This concept had been used before in Boeing transport fiberglass control surfaces. A minimum-thickness, solid laminate skin, supported by closely spaced ribs, was considered in concept 2. This design had been developed in composite control surfaces. Concepts 3 and 4 used the same number of ribs as the existing metal elevator, with solid laminate skin panels stabilized by bead and blade stiffeners. These two configurations were considered because they closely matched the existing aluminum elevator configuration.

The primary factors considered during selection of the composite elevator

206 ADVANCED THERMOSET COMPOSITES

Figure 6-12. Boeing 727 advanced composite elevator.

Concept	Rib Ratio	Fastener Ratio	Weight Ratio	Recurring Cost Ratio	Remarks
1: Minimum-rib honeycomb panel design	1.0	1.0	1.0	1.0	Simple panel tools CHOSEN CONCEPT
2: Multirib unstiffened panel design	8.25	2.5	1.3–1.5	2.6	8 times the number of rib tools
3: Multirib bead-stiffened panel design	3.5	1.5	1.1–1.3	1.7	3.5 times the number of rib tools More complex panel tools Difficult to cocure panels
4: Multirib blade-stiffened panel design	3.5	1.5	1.2–1.4	1.6	3.5 times the number of rib tools More complex panel tools

Figure 6-13. Composite elevator design study summary.

configuration were weight, number of parts required for an elevator assembly, recurring assembly costs, tooling costs, and processing complexity. The results of the evaluation are summarized by concept in Figure 6-13.

Concepts 3 or 4 would have had approximately the same number of parts as the existing aluminum elevator (600), of which 370 would have been carbon-epoxy. The number of parts used in concept 2 would have been even greater than in concepts 3 or 4. Concept one had the smallest number of parts (389), of which 142 were carbon-epoxy. The number of fasteners required to assemble the elevators would have been greater for the concepts with more detail parts in the assembly. The number of fasteners in concept 1 was 1355, or 35% of the fasteners used in the existing aluminum structure. Since concepts 2, 3, and 4 had more detail parts and fasteners than concept 1, they would have been more expensive to fabricate than concept 1. In addition, the bead and blade stiffeners used in concepts 3 and 4 required significantly more processing complexity than the honeycomb-stiffened panels used in concept 1.

The composite elevator configuration selected (concept one) is shown in Figure 6-14. This design features one-piece, lightweight honeycomb upper and lower skin panels, a laminate front spar spliced at the actuator fitting, a laminate rear spar from the elevator tab inboard, and a minimum number of honeycomb ribs.

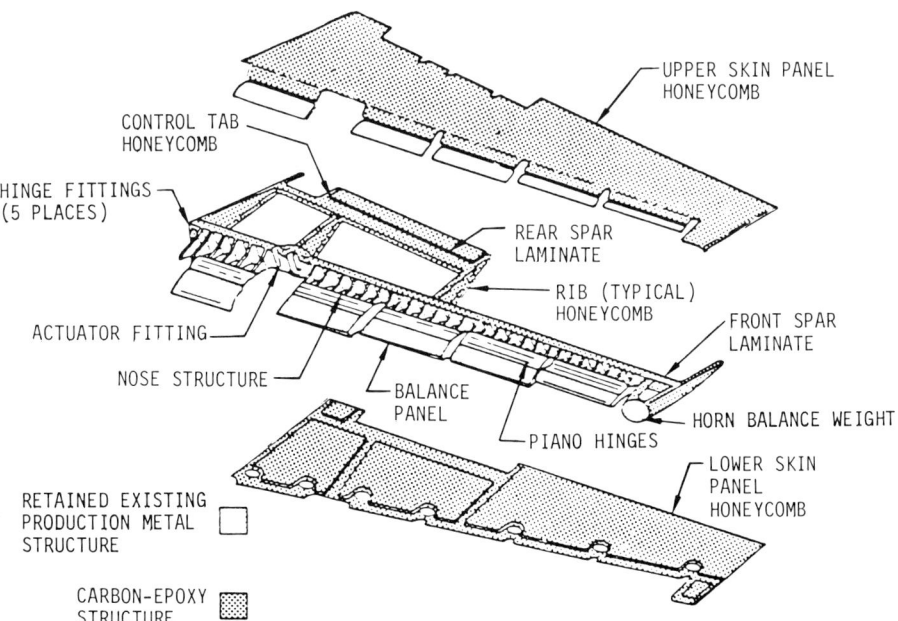

Figure 6-14. Boeing 727 elevator structural arrangement.

208 ADVANCED THERMOSET COMPOSITES

The evaluation results and the concept selected are consistent with the trends of recent Boeing commercial aircraft, where fiberglass honeycomb panels are used extensively on lightly loaded control surfaces, including elevators. The selected design is a refinement of an earlier concept that used existing hinge fittings and had ribs at each elevator hinge. A study showed that redesign of the hinge fitting to introduce the overturning moment directly into the skin panels, combined with the triangular shape of the elevator box, eliminated the need to distribute torsion loads to the panels by these ribs; therefore, the hinge fitting ribs were eliminated.

Another major component redesigned for composites under the NASA ACEE program is the 737 horizontal stabilizer. The stabilizer consists of a structural box, leading edge, tip, fixed trailing edge, elevator, and body gap covers. Figure 6-15 shows the inspar structural arrangement. The structural box was redesinged using carbon-epoxy composite material. To eliminate induced thermal stresses in the trailing-edge panel structure, the trailing-edge beam also was redesigned in carbon-epoxy. The removable portions of the leading edge, trailing edge, tip, and elevator were retained unchanged to reduce program costs. The other components were modified as required to interface with the carbon-epoxy components.

The composite stabilizer was required to be geometrically and functionally

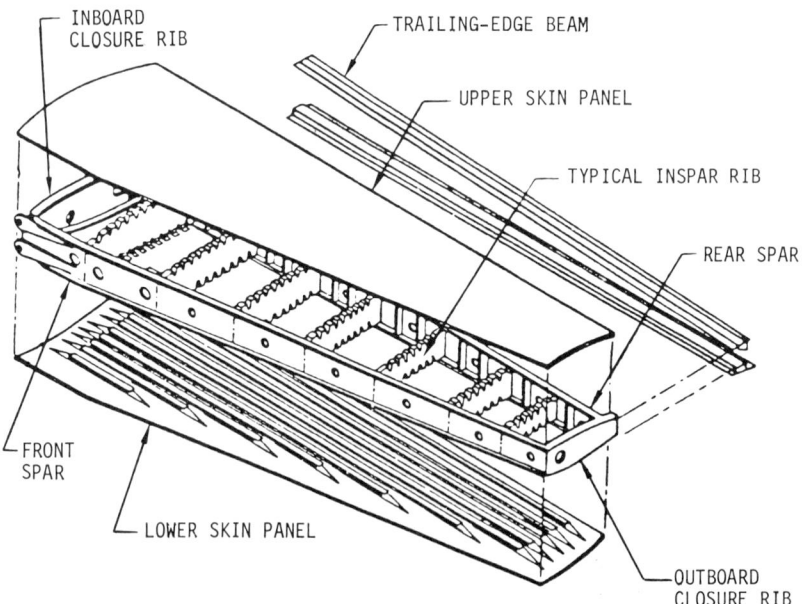

Figure 6-15. Boeing 737 stabilizer inspar structural arrangement.

interchangeable with the metal stabilizer on any 737 without altering the airplane's flight characteristics. In addition, requirements were established to provide protection from any environmental or atmospheric effects peculiar to composite materials. To ensure that the airlines would be able to service the composite components, procedures to maintain and repair them were developed.

The structural body of the metal stabilizer is a multirib design. The skins are made from aluminum sheet stock with bonded doublers at the spar and rib locations. The spars and ribs are built-up aluminum construction using extruded caps and angle-stiffened webs. Each stabilizer is attached to an aluminum center section structure with three bolts at the rear spar and two bolts at the front spar. The third joint on the rear spar is for fail-safety. These five-pin joints are points of interchangeability for the entire stabilizer assembly. The composite stabilizer was required to match this interface.

The structural arrangement selected for the composite stabilizer uses a cocured, integrally stiffened skin, and laminate front and rear spars. The skin is supported on seven inspar honeycomb ribs, and laminate inboard and outboard closure ribs. The trailing-edge beam is laminate construction.

Each skin panel is a single-piece, cocured carbon-epoxy laminate assembly. The basic skin is selectively reinforced with fabric and tape doubler plies. Doublers are provided in the skin at the inboard end where the stiffener and skin loads are transferred out to the spar chords. Doublers also are provided along the spar interfaces to improve the bearing load capability of the mechanical joints. Skin pads and tape doublers are provided at each rib interface to improve the pressure load transfer of the skin-to-rib joint. The skin is stiffened with cocured, I-shaped stringers. The stringers are back-to-back fabric channels with tape caps and are machined after cure to obtain the final cap width and runout.

The rear spar is an I-section that provides attachment flanges for the stabilizer box skins and the trailing-edge panels. The I-section is fabricated by bonding the precured channels back-to-back. Then, constant-thickness precured caps are bonded to this assembly. The spar web is stiffened by mechanically fastened angles. The chord areas increase significantly near the inboard end, where the skin panel loads are transferred into the spar and to the lugs. The spar chords in these areas, where very thick laminates are required, are made with stacked, precured laminate strips. The lug ends are reinforced with solution-treated and aged titanium straps. Access and inspection holes are provided in the spar web.

Two titanium reinforecment straps are required for each lug end. The straps are mechanically fastened to the spar with Hi-Lok®* titanium bolts. Two corrosion-resistant steel flange bushings are press fit into the lug. The bushings are

---

*Registered trademark of Hi-Shear Corporation.

faced and reamed in a master tool after stabilizer assembly. This procedure ensures interchangeability with the center section.

The inspar ribs are a cocured, honeycomb-stiffened laminate. The attach flanges for the front spar and the upper and lower skins are integral with the rib. Protection against moisture absorption is provided by an organic film bonded on the nontool side and by paint on the tool side. The ribs are mechanically attached to the surrounding structure.

The inboard closure rib is a mechanical assembly of constant-thickness, laminated carbon-epoxy elements and current-production aluminum fitting components. The carbon-epoxy elements consists of the rib, upper and lower rib caps, rib channel, and rib stiffeners. The alumninum components consist of five tension fittings and two attach angles. Large access holes are provided in the rib web to facilitate assembly of the closure skin and inspection of the stabilizer internal structure.

The 737 advanced composite stabilizer is 22% lighter that the 737 metal stabilizer and is the first certified commercial application of composites on aircraft primary structure. Five shipsets were manufactured for flight service; installation occurred early in 1984.

NEW-GENERATION AIRCRAFT

Composite components have been used extensively on the new aircraft introduced in the 1980s. Composite applications include control surfaces, fairings, engine nacelle components, interior components, and several others. Over 3000 pounds of advanced composites are incorporated into large aircraft such as the 757, 767, and A310. Smaller aircraft, such as the 737-300, use approximately half this much material. These composite applications are summarized in Figure 6-16.

Some typical applications are shown in Figure 6-17. In general, these components are sandwich constructions. Many consist of carbon fiber-reinforced epoxy skins, with aramid or glass hybridized with the carbon fibers. The design requirements generally are for stiffness and surface smoothness. Smoothness on the outer surface is required for aerodynamic reasons and is created by curing the component against a hard tool. A good example of such a component is the A310 main landing gear hinged fairing door (Figure 6-18), which consists of carbon fiber and aramid fiber face sheets and Nomex honeycomb. The part details are typical for the fixed trailing-edge components on the wings or empennage surfaces of a modern jet aircraft.

Additional all-aramid applications on wing structures include the flap track fairings of the A310 and 767 aircraft. These composite fairings replace glass fiber sandwich construction, which is typical of 747-era aircraft. The design intent in this case was to achieve maximum weight reduction at minimum cost.

	Airplane				
Component	MD-80	737-300	A310	757	767
Main landing gear doors		N/A	H2	H2	H2
Nose landing gear doors				H2	H2
Wing-to-body fairings	A2		A2	H2	H2
Flap support fairings			A2	H2^a	H2
Inboard ailerons	N/A	N/A	N/A	N/A	C1
Outboard ailerons	C2	C2	C2^a	C2	C2
Spoilers	C2	C2	C3-inner C1-outer	C2	C2
Nacelle cowl components	H2	H2	H2	C2	H2
Wing fixed leading-edge panels				H2	
Engine strut fairings		A2	A2	A2/C2	A2
Elevators		C1		C1	C1
Rudders	C1^a	C1	H1	C1	C1
Horizontal stabilizer structural box		C3			
Horizontal stabilizer fixed trailing-edge panels				H2^a	H2
Vertical stabilizer fixed trailing-edge panels				H2	H2
Stabilizer and fin tip fairings		H2	H2	A2	A2
Wing fixed trailing-edge panels	A2			H2	H2
Wing trailing-edge flap components				H2^a	A2
Cargo compartment liners				A3^a	A3
ECS ducting			A2	A2/3^a	A2/3^a
Floor beams					
Center wing rear vapor barrier				H2	
Cargo compartment floor panels					A2^a
Stowage bin components					A3
Lavatory module					A2
Off-wing escape slide compartment door					H2
Floor panels			C2		

Figure 6-16. Increasing use of advanced composites: A = aramid-epoxy, C = carbon-epoxy, H = hybrid (combination of aramid and carbon, or fiberglass and carbon); 1 = honeycomb components, 2 = full-depth honeycomb, 3 = solid laminate, a = under study.

212 ADVANCED THERMOSET COMPOSITES

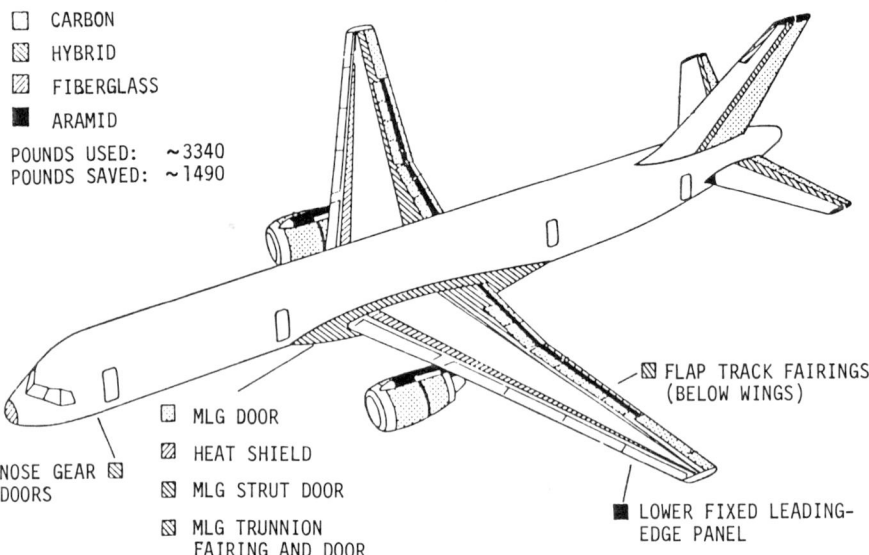

Figure 6-17. Boeing 757 composite applications.

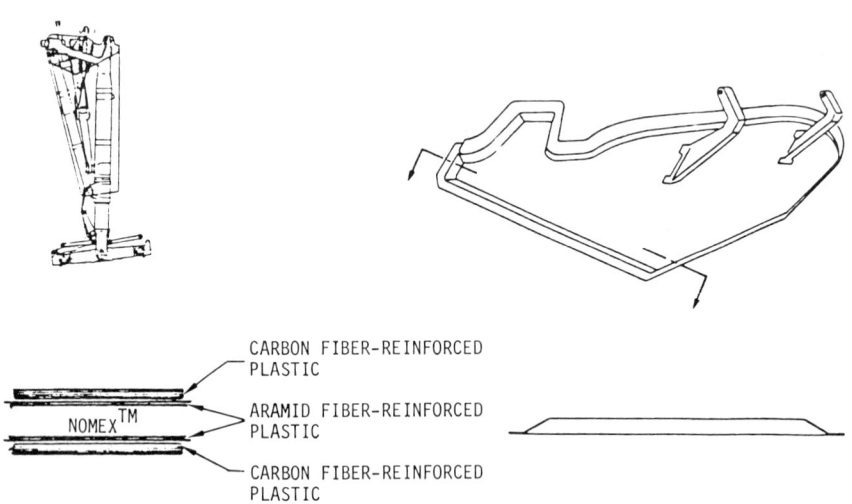

Figure 6-18. Airbus Industries A310 main landing gear hinged fairing door.

COMMERCIAL AIRCRAFT APPLICATIONS   213

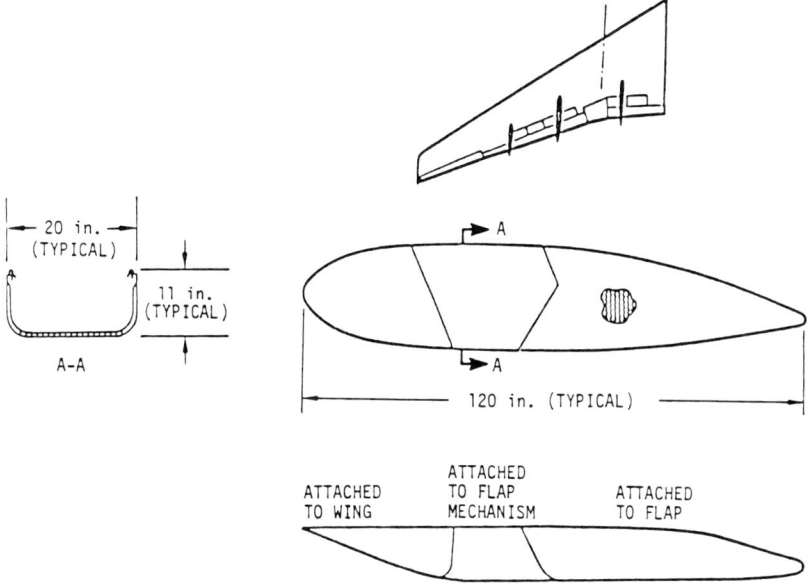

Figure 6-19. Boeing 767 flap linkage fairings.

On the 767, the sandwich skins are aramid fabrics cocured to form a sandwich with Nomex honeycomb core. The A310 uses the same general concept and achieves an additional cost saving through the use of aluminum honeycomb core. Details of this type of construction are shown in Figure 6-19.

All control surfaces of the new-generation aircraft use composites to some degree. Control surfaces include such items as the spoilers, ailerons, flaps, and air brakes. An example of a full-depth honeycomb application is the 767 aileron (Figure 6-20). This composite article is nearly 19 feet long and is fabricated of carbon skins and a carbon spar, bonded to Nomex honeycomb. In this application, the skins and spar are precured to the final shapes and dimensions, then bonded to the honeycomb core using conventional adhesive bonding technology. One of the reasons for precured skins is that both the upper and the lower surfaces of the aileron must be aerodynamically smooth. Precuring the skins results in a more economical tooling concept than attempting to capture and cure the entire part in a single-stage operation.

Another example of a control surface manufactured using a full-depth honeycomb core design is the 767 carbon fiber-reinforced spoiler (Figure 6-21). The spoilers, located on the outer wing, are designed for frequent use during takeoff and landing. The design cosists of a sandwich construction with carbon-epoxy top skins, bottom shell, and side ribs bonded to Nomex honeycomb core of triangular cross section. Skins and side ribs are made of seven plies of uni-

214 ADVANCED THERMOSET COMPOSITES

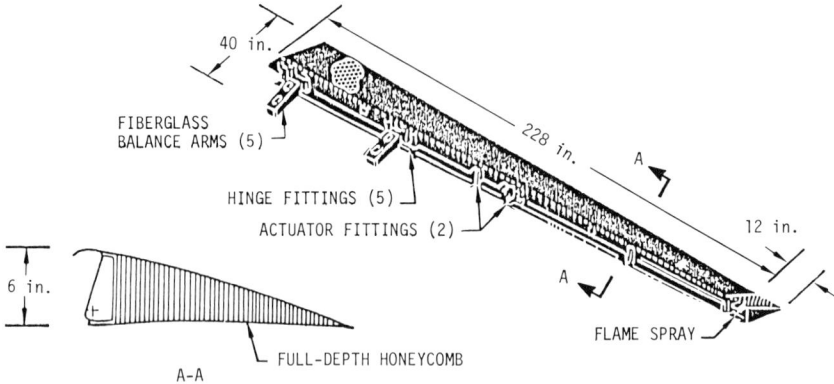

Figure 6-20. Boeing 767 outboard aileron.

directional preimpregnated carbon tapes, with local reinforcement in the actuator area. A sandwich construction was chosen for the spoilers primarily because of the unsymmetrical loading of the spoiler surfaces, especially in the spanwise direction. In this concept, manufacturing begins by laying up the lower shell laminate and adhesively bonding it to the Nomex honeycomb core. The honeycomb core is machined, and the spoiler subassembly is bonded to the spar. Then the honeycomb core is finish milled, and the upper laminated skin is bonded to the remaining structure. A very similar concept is used on the 757.

A different approach was used by Airbus for the inner and outer air brakes [6]. A monolithic structural concept was used to meet the design objective of a

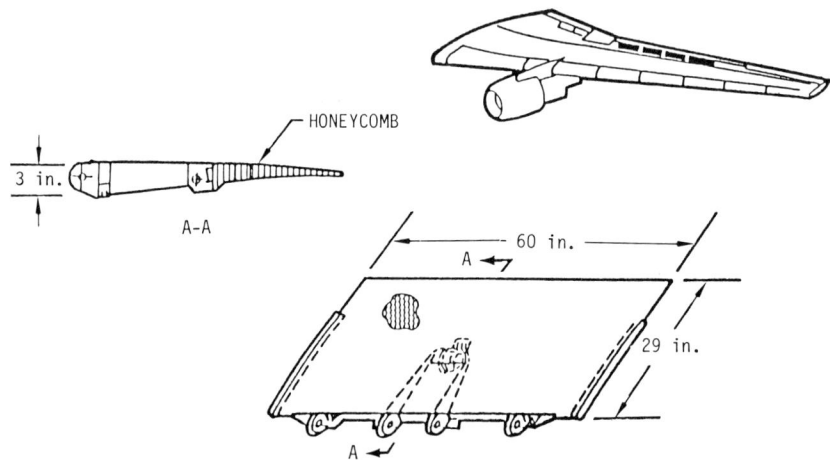

Figure 6-21. Boeing 767 outboard spoilers.

light, rigid, structure that would withstand the flexural and torsional deformation induced by high compression on the outer skin. In this instance, the skins and ribs are made of unidirectional preimpregnated carbon fiber tapes, with varying number and orientation plies. The skin and rib assembly is cured in an economical, single-stage operation. The metal front spar and integral machine hinges and brackets are bolted to the finished carbon fiber-reinforced component.

Skin plies vary from 26 at the actuator attachment to 8 at the trailing edge. The construction, strikingly different from honeycomb core designs, is shown in Figure 6-22. A weight saving of approximately 70 pounds per aircraft is achieved on the A310 using this design concept.

The empennage areas of the 767, 757, A310, and 737-300 also are extensively designed in advanced composites e.g., the elevator and rudders. The fixed trailing-edge assemblies, as well as the thin leading edge of at least one on these aircraft, also are made of advanced composites.

In terms of total span, the 767 rudder is one of the largest composite components in production today. Nearly 36 feet long at the front spar, this sandwich panel design uses carbon fiber fabric and tape coupled with Nomex honeycomb core. The rudder cover panels, spars, and ribs are all of honeycomb sandwich design. The sandwich panels are bonded in one operation. Some of the assembly details are shown in Figure 6-23. The side panels or cover panels are bolted together at the trailing edge using special tubular titanium rivets. These components are significantly lighter and offer other design advantages in terms of

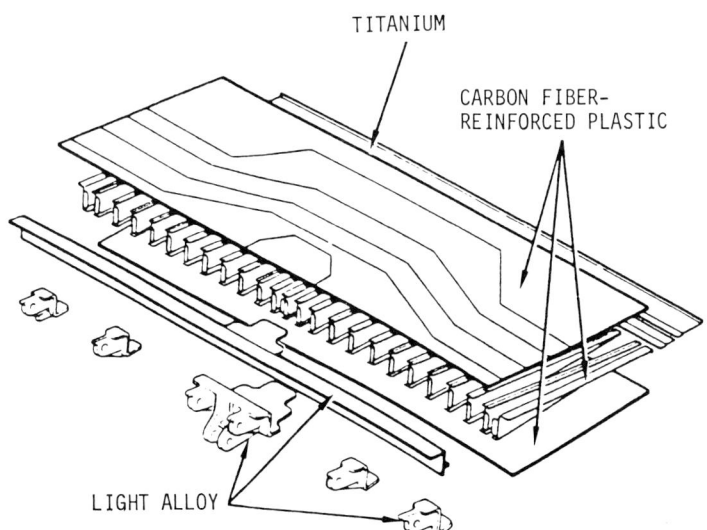

Figure 6-22. Airbus Industries A310 inner air brakes.

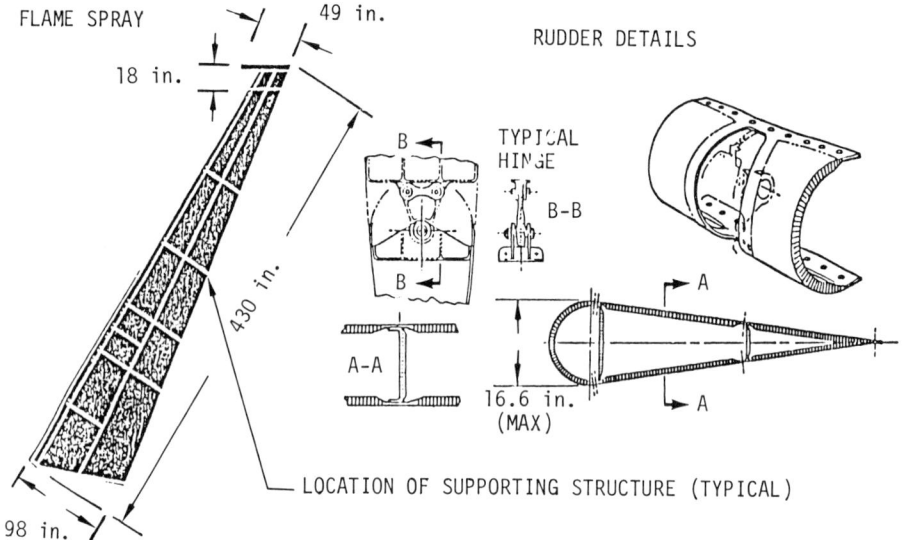

Figure 6-23. Boeing 767 rudder.

total system stiffness over metal components. In addition to the carbon fiber, some fiberglass or glass-reinforced materials are used, particularly in areas where lightning protection systems have been applied. Restricted to tips and leading-edge areas, lightning-strike protection generally includes protective foils or metal sprays to dissipate electrical charges that develop during lightning strike.

The engine pylon fairings provide aerodynamic continuity between the wing lower surface and the engine pylon for optimum air flow at the engine-wing interface. These elements are subjected to high-frequency vibration, extremely high noise intensities and airflows, and occasionally, elevated-temperature effects. The fairings are generally manufactured using aramid fabric face sheets and honeycomb core. These components are very similar to the wing-fuselage fairings. The wing-fuselage fairings are aerodynamic sealers between the wing and the fuselage; they must be resistant to aerodynamic forces and yet flexible enough to be compatible with the deformation that occurs between the fuselage and the wing. Some designs of these components use all-aramid fabric with Nomex honeycomb core. Other designs use carbon fiber fabrics over the honeycomb core bay areas, as well as aramid fabrics in the honeycomb bay and the laminate edgebands. These large fairings generally save substantial percentages of weight compared to the glass fiber constructions of earlier aircraft. The use of both carbon and aramid fibers in these components reduces weight from that of all-aramid or all-glass designs.

Composites also have made inroads into the design of engine nacelles. In the modern jet engine, the inlet cowls, fan cowls, and translating cowls are made of carbon fiber, aramid fiber, or hybrids of aramid and carbon fiber. On some engines, these designs use Nomex honeycomb core and carbon fiber materials. Others use aluminum honeycomb to take advantage of the greater stiffness of aluminum, as well as the additional thermal conductivity of the aluminum honeycomb which allows the skins to operate at lower temperature. Typical examples of advanced composite desgin applications for engine nacelles are shown in Figure 6-24.

Generally, the engine nacelle areas require elevated-temperature-curing resin systems, not only for structural durability, but also because of temperature requirements imposed on the nacelle during normal operating conditions. These requirements do not apply to many of the fixed trailing-edge, wing-body fairing or door applications, where lower-temperature operating conditions exist. Stiffness or strength requirements at elevated temperatures are not nearly as severe, and lower-temperature-curing resin systems are commonly employed.

One of the advantages of early glass-reinforced resin structures, particularly in damage-prone areas, was ease of repair. This technology has progressed to cover the repair of advanced composite structures. Repairs are simple and are made with room-temperature- and elevated-temperature-curing materials. Heating blankets and vacuum bags are commonly used. The repair materials, temperatures, and methods are functions of the component design and the damage

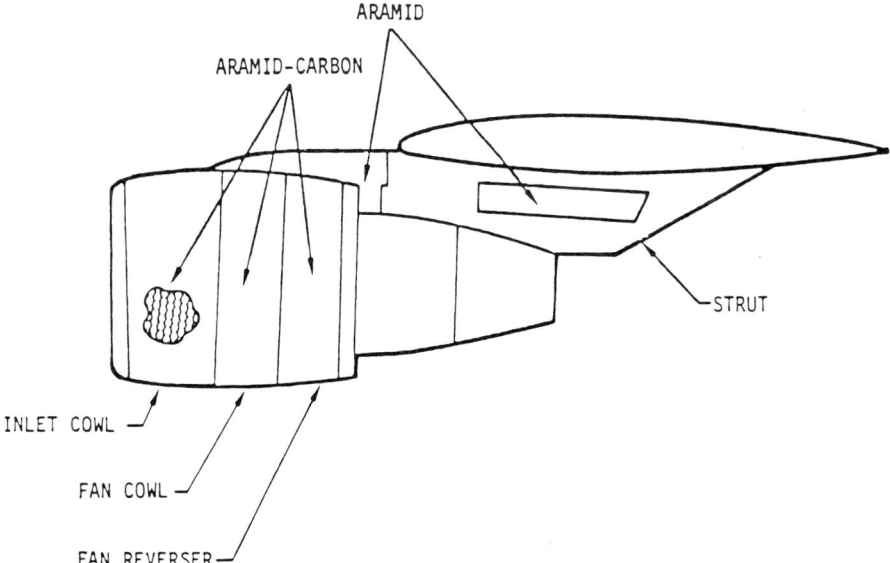

Figure 6-24. Boeing 767 engine strut/pod.

218 ADVANCED THERMOSET COMPOSITES

type, severity, and location. Information on repair procedures for composite structures is incorporated directly into structural repair manuals for each aircraft. The documents provide a section on general damage evaluation and repair, as well as detailed information for each individual component. Each component is shown schematically, and a table is included showing damage-critical areas and repair procedures. Some components and some damage types in critical areas require specially engineered repairs. In these cases, the component generally is removed and returned to the manufacturer for disposition. A schematic of a common repair procedure is shown in Figure 6-25.

Because of their very attractive structural and manufacturing properties composite materials are widely used in the interiors of commercial aircraft. In addition to meeting mechanical property and fabricability requirements, all materials used within the pressurized envelope of the aircraft must meet both the flammability resistance requirements defined by regulatory agencies and the additional flammability and smoke and toxic gas emission guidelines of the manufacturers. Additionally, visible portions of the cabin and interior components must meet high aesthetic requirements to satisfy the airlines and their customers. Durability and maintainability also are important considerations for in-

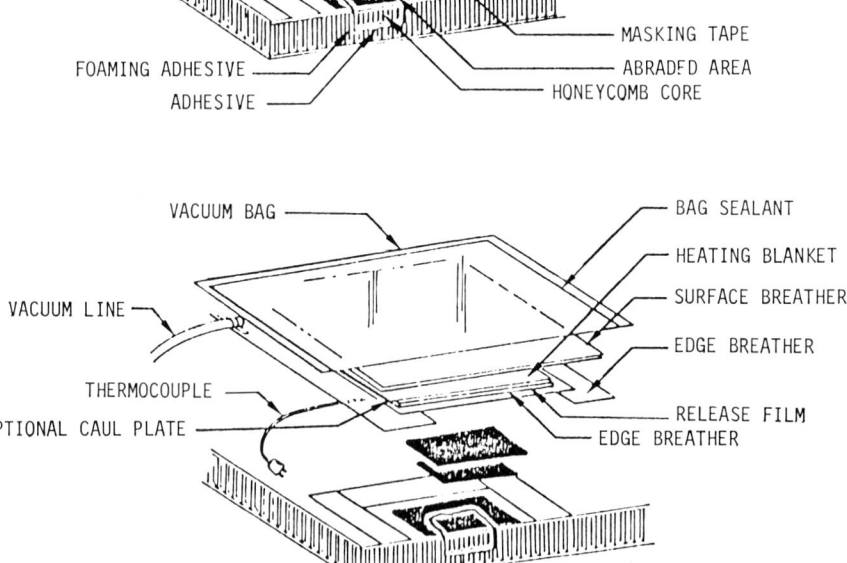

Figure 6-25. Advanced composite permanent repairs lay-up/cure.

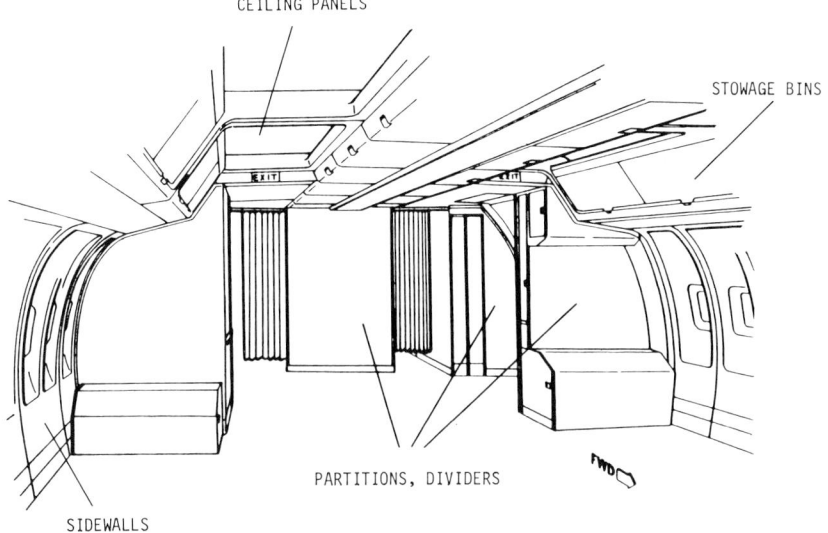

Figure 6-26. Typical aircraft interior.

terior applications. Composite materials play an important role in meeting the diverse requirements of commercial aircraft interiors.

Figure 6-26 shows a schematic of a typical new-generation aircraft interior. The majority of the parts are fiberglass. In many of the applications, such as sidewall and ceiling panels, the resin system is phenolic. This system is used widely because of its excellent fire properties, including low flammability and low smoke and toxic gas emissions. The phenolic systems available today do not have the physical and mechanical properties required for all interior composite applications or for use with all fibers. The other commonly used resin systems are epoxies.

Schematics of two representative sidewall panel construction methods are shown in Figure 6-27. The crushed-core sidewall panel is used in large areas of the interior, much of which has decorative surfaces. The crushed-core concept was developed for use with the phenolic resin systems because of the relatively poor adhesive properties of the phenolic. The increased bonding surface area of the crushed core compensates for the low phenolic properties. A more standard use of honeycomb is the conventional expanded-core panel. These very light panels are decorative and, with decorative layers on both sides, may be used as partitions.

Several different design concepts have been used for aircraft passenger floor panels. Some of the more recent designs employ a carbon-epoxy honeycomb sandwich concept. Similar constructions use fiberglass in place of the carbon.

220 ADVANCED THERMOSET COMPOSITES

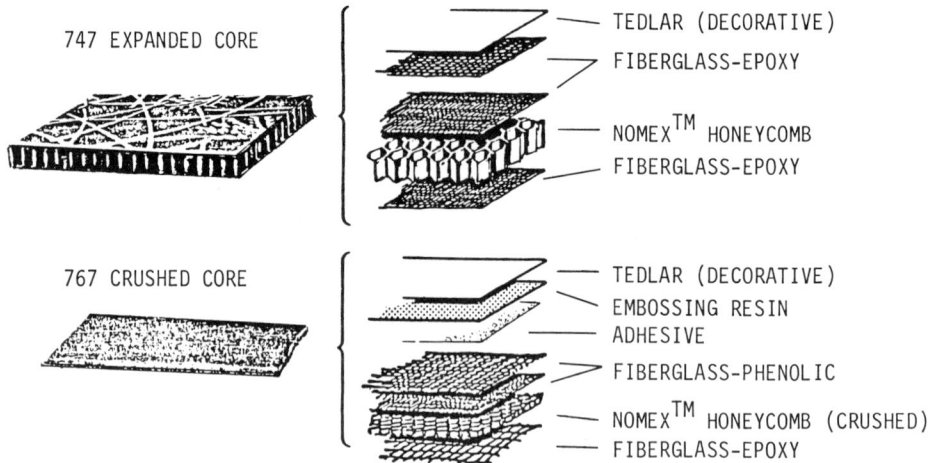

Figure 6-27. Sidewall panel construction.

Other composite, honeycomb-reinforced parts include stowage bins, ceiling panels, cargo liners, and lavatory modules.

Other interior applications that take advantage of composite capabilities include low-temperature air-distribution ducts, which are laid-up fabric; filament-wound tanks and torque tubes; and springs.

Three components located in the 767 cargo hold area shown in Figure 6-28. In the background is an air-distribution duct. These low-pressure air-distribution systems are used extensively throughout the airplane. Laid-up aramid fiber fabric ducts provide a very lightweight system. The potable water tank is an aramid fiber, filament-wound tank. This tank is part of a pressurized water system and is designed to take advantage of the excellent specific tensile strength properties of the aramid fibers. By contrast, the lavatory waste tank is a carbon fiber, filament-wound tank. This tank is part of the vacuum waste system, and use of carbon fibers reflects carbon's superior stiffness and compression strength properties.

Another carbon fiber, filament-wound part is the stowage bin torque tube (Figure 6-29). In both the tanks and the torque tube, the fittings are metal.

A filament-wound composite door spring (Figure 6-30) has been developed for the 767. As a result of using unidirectional graphite fibers in an epoxy matrix, the springs are one-third as heavy as comparable steel springs and only half the weight of state-of-the-art titanium door spings. The spring consists of graphite fibers contained in a rectangular cross section, looped to form the cylindrical shape. The linear length of the uncoiled filament wound composite tube is 800 inches long. The spring consists of 26 coils and it is 14 inches in diameter. Weight savings are over 100 pounds per aircraft.

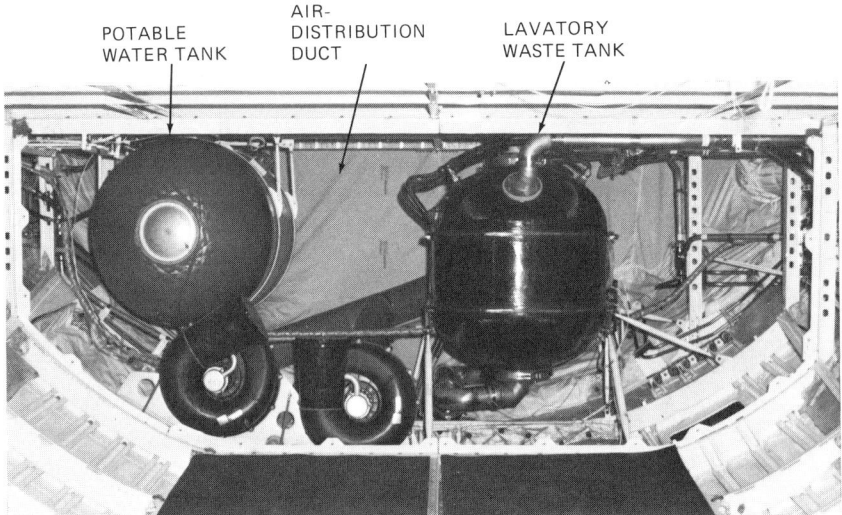

Figure 6-28. Boeing 767 cargo hold—composite components.

THE FUTURE

The future of the commercial aircraft industry appears bright. The jet age began in earnest in 1958, and with an economic life of approximately 20 years, all the commercial jet aircraft produced to date will be replaced over the next 20 years or so. In addition, with a projected worldwide traffic growth of about 5% per year, the passenger-carrying capacity of the air transport fleet will need to double over the same time period. Coupled with a projected increase in cargo requirements, the forecast appears rosy.

There are dark clouds, however—particularly the economic health of the world's airlines. The need for new airplanes is real; the question is, Can the airlines afford the new equipment? From this quandary, the airframe manufacturers are engaged in the high-stakes "sporty game" [7] of trying to anticipate the needs of the marketplace. Portions of the equation include airplane size, type, range, payload, number and type of engines, and operating economics.

One thing is abundantly clear: airlines need highly efficient aircraft at a price they can afford. The role of advanced composites in the future has not yet totally crystallized. Composites potentially can make a major positive impact on aircraft range, payload, and operating economics. Composite structure can be substantially lighter than conventional aluminum structure; additionally, its excellent strength and stiffness properties and ease of fabrication allow improved aerodynamic and structural efficiencies in what could be considered radical new designs.

222 ADVANCED THERMOSET COMPOSITES

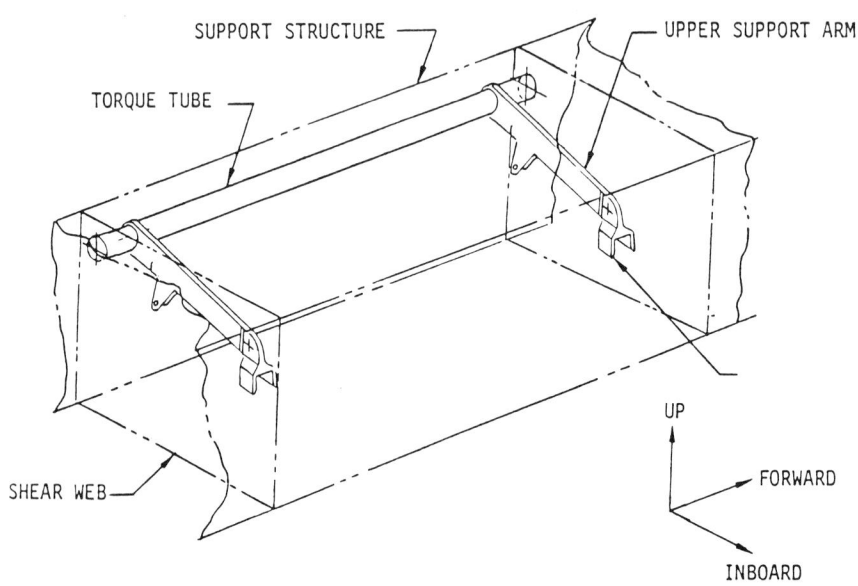

CENTER BIN--TORQUE TUBE INSTALLATION

Figure 6-29. Stowage bin filament-wound torque tubes.

Figure 6-30. Carbon fiber–epoxy spring.

Operating economic considerations are strongly influenced by fuel price. After a long period of price stability, jet fuel prices over the last decade have dramatically increased. Prices in early 1973 were approximately 10¢ per U.S. gallon; following the Middle East crises of 1973 and 1977, and the Iran crisis of 1979, prices peaked at an average of over $1.00 per gallon in 1981. Relative stability and even modest price decreases have followed, with the 1983 world price averaging approximately 92¢ per gallon.

The effect of fuel price increases on direct operating costs (DOC) for the airlines may be seen by comparing the DOC breakdown in early 1973 with that in 1980. Figure 6-31 shows these data for a large commercial jet aircraft.

The last few years have brought some uncertainty in jet fuel prices; however, it is generally anticipated that in the long run fuel prices will continue to rise. At current price levels and in anticipation of higher fuel costs, there will be continued customer demand for airplanes with improved fuel efficiency. These improvements may be achieved in several ways, including improved engine performance, improved aerodynamic efficiency, and reduced aircraft weight. Composites have a major role to play in the last two categories.

Projections of material usage distributions are shown in Figure 6-32. The projections for composite usage extend from the current 3% to 25–65% of the

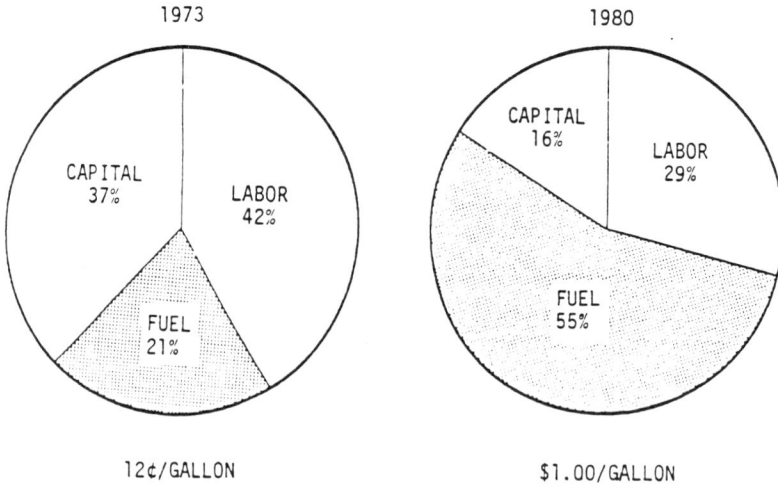

Figure 6-31. Direct operating cost (DOC) elements (1000 NMi trip).

aircraft structural weight. On one fact, all forecasters agree: composite usage will increase. The question is how much.

Current composite applications are in high-surface-area, low-specific-density parts, primarily in secondary structure. Little expansion can be achieved in this area because the conversion is already nearly complete. In the future, composite usage will expand into the high-specific-density parts of primary structure. These can be divided into three broad areas: the tail section, the fuselage, and the wing. The usage projections differ based on which broad areas are converted, since each has its own design and fabrication problems, and each could be converted separately while the others remained metal.

In ranking current design and fabrication capability, the tail section is the most advanced, following certification of the 737 horizontal stabilizer. Progress in wing technology is encouraging, with production commitments anticipated within the next decade. Large commercial aircraft fuselage technology is still in its infancy. The materials incorporated in each area will be strongly influenced by the timing of the decision relative to design and fabrication readiness, and whether the aircraft is a design improvement, a derivative, or a totally new design.

A design improvement may be considered where there are major airplane performance or manufacturing gains possible. These could include replacement of certain complex, expensive metal parts with more easily fabricated, lighter composite parts or incorporation of entirely new designs if the weight savings are significant and the costs acceptable.

For a derivative aircraft, there is a strong motivation to use as many of the original components as possible to take advantage of the reduced tooling costs.

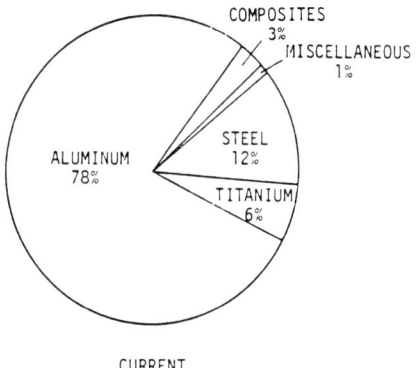
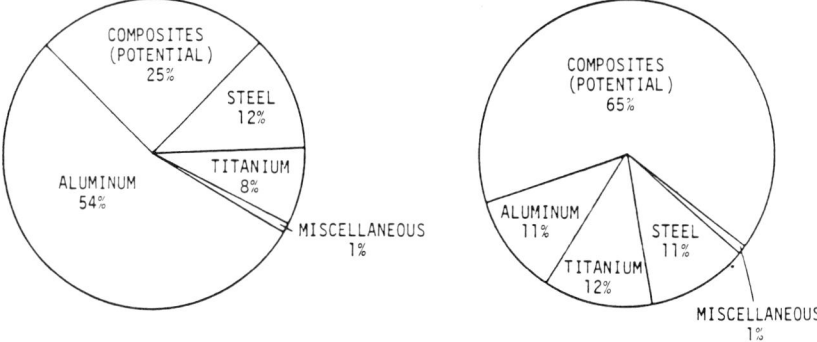

Figure 6-32. Advanced technology airplane structural materials weight distribution: 1985–1995 subsonic airplane.

Thus, composites will most likely be incorporated into those components that are being redesigned for other reasons.

The optimum use of composites requires a totally new design. This allows the freedom to take maximum advantage of composite properties, while at the same time allowing the resizing and tailoring of the entire airplane for the specific range/payload mission.

For aerodynamic reasons, a high-aspect-ratio wing is desirable. The strength- and stiffness-to-weight properties of advanced composites allow more freedom to design a high-aspect-ratio wing. With the design and production flexibility of composites, future designs may be substantially different from current commercial aircraft. Some of these new design and fabrication concepts are already showing up in smaller civil aircraft, such as the all-composite Lear Avia, the Avtek 400, and the Beech Starship I.

The seeds of the future have already been planted. The NASA programs—beginning with secondary structure, and followed by wing and fuselage studies—are laying the groundwork for the aircraft of the future. Material suppliers are developing improved material systems. Airplane fabricators are evolving the design concepts, automating and refining the manufacturing methods, and developing the associated technology required to design and build the all-composite aircraft. Additionally, the groundwork for certifying the all-composite aircraft is being laid by the NASA programs, the current secondary structures usage, and the general aviation market.

References

1. Berg, M. *1982-83 Graphite Prepreg and Fiber Market Survey Report*, Composite Market Reports, Inc., San Diego, Calif. advance copy Oct. 22, 1982.
2. Poulos, N. E., "Historical Development of Radomes" Chapter 1, *Radome Engineering Handbook*, Ed. by J. B. Walton, Marcel Dekker, New York, 1970, p. 1.
3. Wood, Richard, "Glass Reinforced Plastics in the Aircraft Industry" Chapter 18, *Glass Reinforced Plastics*, Ed. by P. Morgan, Illif, London, 1961, pp. 267-270.
4. Thompson, Vere S., Composite Applications on Boeing Commercial Aircraft, *Proceedings of the 26th National SAMPE Symposium*, April 28-30, 1981.
5. Stone, R. H., "Flight Service Evaluation of Kevlar-49 Epoxy Composite Panels in Wide-bodied Commercial Transport Aircraft," NASA Contractor Report 172344, Lockheed-California Company, Burbank, California, June 1984.
6. Koshorst, J. "Advanced Composite Structures in Commercial Transport Aircraft," *Kevlar in Aircraft, Summary—Technical Symposium IV*, E. I. du Pont de Nemours & Co., Geneva, Switzerland, October 1982.
7. Newhouse, J., *The Sporty Game*, Alfred A. Knopf, New York, 1982.

7
PROCESS INDUSTRY APPLICATIONS

Joseph S. McDermott
President, Composites Services Corp.

The uses of composites in processing environments are among their most tested, yet least publicized applications. Where chemicals or other inherently corrosive materials are handled, or where the by-products of converting organic materials create harsh fluid or vapor environments, glass-fiber-reinforced plastics have been proven to have major advantages over wood, glass, and metallic container or construction materials. This fact has two significant effects on the approach taken in this chapter.

First, the critical nature of tanks, pressure vessels, piping systems, ductwork, and other industrial process equipment has made it necessary that each component be thoroughly evaluated through standardization, testing, and certification procedures. Thus, generalizations must be made cautiously. What may apply to an industrial coating within a pulp mill fume stack, for example, may not apply to equipment designed to carry loads and be used with automatic sprinkler systems as an element of a site's fire safety. The treatment of applications of composites in this chapter will, therefore, be quite specific as to product and use.

Second, because industrial equipment is large in size and represents heavy investment, it is not a prime field for experimentation in regard to performance. Hence, the composites used in process industry applications tend not to be aramid, graphite, or other sophisticated reinforcement systems in an unusual polymer matrix. Notable exceptions will be discussed. Unless otherwise specified, therefore, the reference to composites in this chapter is to glass-fiber-reinforced polyester (FRP) or epoxy components, best known in the industry as "fiberglass" tanks, piping systems, grating, ductwork, etc. It is these materials, in special formulations for the corrosives they will handle or the particular environments of their use, that are appropriate to process equipment. Certain physical and mechanical properties and aesthetic characteristics required for aerospace, automotive, and other applications of advanced composites are not usually pertinent for this equipment.

Terminology in this chapter is therefore contextual. Where codes, standards, and reference literature are cited, their use of generic terms and short forms such as "RTRP" is unavoidable. The chapter avoids use of "FRP," however, to avoid confusion with "fire-retardant plastics." The preferred term of industrial equipment users—"fiberglass,"—is employed except where the reinforcement may not be glass fiber or where "reinforced plastics," "RP," or composites is simply the correct nuance.

PETROLEUM AND CHEMICAL INDUSTRIES

Although it is feasible to group petroleum and chemical processing applications because of their close relationship, in many cases their uses of composites are separable and invite distinct treatment. This fact becomes clear as standards, testing, and limitations are discussed. Furthermore, there are certain applications—such as the introduction of fiberglass sucker rods in the oil patch—that the supplying industry records as an "energy industry application" rather than a petroleum processing use.

Storage Tanks

Fiberglass equipment for holding corrosive materials typically consists of several layers: (1) an inner resin-rich surface designed to withstand the intended corrosive contents; (2) an interior transition layer, also relatively resin rich, and reinforced with chopped glass; (3) a filament-wound layer of continuous strand or woven roving impregnated with resin for structural integrity; and (4) an exterior resin-rich surface for added resistance to spillage in the corrosive environment. When a mandrel is not used for filament-winding the structural layer of the tank, the laminate can be built up by contact-molding alternate layers of chopped strand and woven roving. In both filament-wound and contact-molded tanks, the resin-rich interior will include a chemical-resistant mat composition or an organic veil.

Glass fiber, although offered in various grades for different applications, is not resistant to strong acids or alkalies and so must be thoroughly "wetted out" by the corrosion-resistant resin. For this reason, glass content in the service layer, which is usually between 10 and 20 mils thick, is reduced to approximately 20 percent by weight.

Surfacing veils are thin, highly porous mats of glass or organic fibers used to produce resin-rich interior surfaces. The optimum surface for chemical resistance would be composed of 100-percent resin, but unreinforced thermoset resins are relatively brittle and crack under the mechanical or thermal stresses normally encountered in use. A surface with good chemical resistance and strength can be made by reinforcing a resin-rich layer with 10-percent fibers of C-glass.

C-glass was developed specifically for chemical resistance. Unfortunately, since the physical properties of this glass prevent the economical production of continuous fibers, it is available only as a highly porous surfacing that is mat 0.010 to 0.030 in. thick. C-glass mat is manufactured with various finishes that make it compatible with the various resins used in the manufacture of corrosion-resistant equipment.

Although C-glass has good resistance to acid solutions, it is attacked to an appreciable extent by hot caustic solutions. This limitation of C-glass mat has led to the development of a synthetic surfacing using C-glass as reinforcing. The improved performance of this surfacing is attributed to the increased elongation (25 vs 7 percent) of the organic fiber and the combination of vertical and horizontal orientation of fibers in the mat. Because of the caustic resistance of synthetic veils, many resin manufacturers now recommend this material as the interior reinforcement for caustic and hypochlorite applications. The diminished availability of C-glass also accounts for growing use of synthetic veils.

E-CR glass, a new chemical-resistant grade of E-glass (originally developed for electrical applications), has recently been introduced on a commercial scale. This glass is reported to be superior to C-glass in acid resistance and superior to E-glass in overall chemical resistance. It can be pulled into continuous filaments and is available in the same forms as E-glass.

The commercially significant resins used in fiberglass storage tanks are polyester and vinyl ester. Furan resins offer surpassing chemical-resistance properties, but they must be handled with special cure schedules and other attention. If these problems can be managed, furan resins should be exploited for their ability to handle strong solvents as well as acids and bases.

The wide range of available polyesters and vinyl esters combine relatively low cost and ease of fabrication. Bisphenol A fumaric acid polyester has good resistance to mineral acids, oxidizing bleach solutions, and salt solutions. Isophthalic polyester resins are generally less costly but lend themselves to formulation over a wider range. Production from more expensive glycols, for example, leads to better cured-resin properties.

Vinyl ester resins are produced when polyesters are built up with epoxy resins, at whose reactive sites a reactive vinyl group is attached. Now the "hybrid" product can be cross-linked with styrene in the manner of other polyesters. Furthermore, the epoxy backbone of this system permits the cured product to elongate under stress and to absorb mechanical and thermal shocks. These vinyl ester resins, giving up none of the chemical resistance of polyesters, are made from liquid bisphenol epoxy resin. Different proprietary raw-material starting points have led to a new generation of high-performance vinyl esters with increased temperature range and solvent resistance.

All of these resins may be further upgraded by introducing chlorine or bromine into the molecular structure. The resulting fire-retardant properties can be enhanced by the addition of antimony trioxide filler.

Design of chemical service equipment refers both to design of the laminate for its particular service and to design of the structural characteristics of the tank at its site. Both are covered in basic fashion in the following standards of the American Society for Testing and Materials [1]:

ASTM D3299 (1981) Standard Specification for Filament Wound Glass Fiber Reinforced Thermoset Resin Chemical Tanks
ASTM D4097 (1982) Standard Specification for Contact Molded Glass Fiber Reinforced Thermoset Resin Chemical Resistant Tanks

These minimum design parameters and construction elements were formerly covered in National Bureau of Standards Voluntary Product Standard PS 15-69. The Bureau no longer sponsors commercial standards of this type, and PS 15-69 was withdrawn in 1980 because the ASTM standards have supplied specifiers' needs for the most part.

The scope of both of these industry standards should be studied carefully. They were written to cover the design and construction of tanks intended to operate under a specific set of conditions. If the conditions of the application under buyer consideration exceed those stated, other special designs will be required; in some cases, specialized construction procedures will be required as well.

If there are no unusual conditions involved in a tank's application, the selection of one of the two industry specifications may be governed by buyer preference or left to the choice of the fabricator. The difference between the two specifications involves the fabrication method used in the structural portion of the tank shell. Since the two methods of fabrication use different reinforcement materials as well, they produce tanks with different physical properties. If properly designed and fabricated, fiberglass equipment may be produced to either industry specification and give comparably satisfactory service life. The choice will be dictated by the specific chemical and physical requirements for a given application. More often, the choice is a matter of fabrication economics or the technology available from a given fabricator. Past buying practices may also play a meaningful role in the decision.

For a purchase description, tank designers will require the following:

1. A complete description of the chemical environment (including concentrations and specific gravities of all chemicals to be contained) and the operating temperature of the environment;
2. Special loading requirements, such as pressures, vacuum conditions, loading from ladders, catwalks, or agitation;
3. Any seismic design requirements and/or wind loads.

A drawing of the accessories and ancillary items that will be part of the tank and its installation conditions is also desirable.

The importance of detailing the complete tank description cannot be overemphasized. Such details will govern construction items such as vent size in relation to the tank's shell design. For example, will the tank be filled by air-loading or by pumping? Will the tank be vented into the atmosphere or into a fume conservation system? These and other considerations are absolutely necessary to proper tank design, and a knowledge of details is necessary to determine the proper classification of the tank under the industry specifications. An eight-page booklet, *Users Guide to RP Industrial Equipment, #1—Tanks*, is available from the Corrosion-Resistant Structures Committee of the industry trade association [2].

The physical design of vertical tanks may include sloped, disked, flat, or conical bottoms. Horizontal tanks, large tanks in excess of 100,000-gallon capacity (constructed in sections and shipped by conventional means), and the special requirements for underground tanks are all beyond the scope of this chapter. Tank manufacturers supply general criteria for each of these conditions, and major chemical producers who use this equipment have accumulated their own in-house specifications from their user experience. A good third-party treatment of design detail can be found in Chap. 5 of *Fiberglass-Reinforced Plastics Deskbook* [3]. In 1981, a joint committee of the RP equipment users from the Materials Technology Institute (MTI) and the Society of the Plastics Industry (SPI) published a *Quality Assurance Report on RTP Corrosion-Resistant Equipment* [4], which also includes design formulas not yet found elsewhere. It is essential to note that this series of recommendations and educational material does not constitute a specification or mandatory code. The publication is intended to propose criteria that will provide a basis, when evaluated over several years, for the code work of the American Society of Mechanical Engineers Standards Committee on RTP Corrosion-Resistant Equipment. This group is scheduled to publish a draft acceptance code in 1986 [5].

Materials Comparison

Reinforced plastics offer many advantages in the construction of corrosion-resistant equipment. As with any material, there are temperature, strength, and chemical-resistance limitations. Reinforced plastics cannot be substituted into a steel design arbitrarily, but if a reinforced plastic is suitable for the application at hand, a considerable economic advantage may be realized. Designers and cost-conscious purchasers must both remember that the RP equipment supplier is much more than a "fabricator" who assembles his product from supplied materials such as sheet metal. As described earlier, *the RP producer creates his own material of construction simultaneously with the physical configuration.*

Fiberglass tanks are usually specified because of the corrosion-resistant requirement. Their initial price is usually less than that of stainless steel or other corrosion-resistant alloys. There are many circumstances, moreover, that can make fiberglass competitive with lower-cost materials such as carbon steel. Savings can result from their ease of installation. Complete tanks have been lifted into place using cranes or helicopters, leading to considerable reduction in installation costs (one-seventh that of steel and one-half that of aluminum). Also, the light weight of RP often allows equipment to be mounted on existing structures, thus leading to additional savings. These economies should not be ignored in the evaluation of a project.

Fiberglass equipment and piping systems are electrically nonconductive and can be used to isolate processes from stray currents or to replace metal components that have failed because of galvanic corrosion. If electrical conductivity is required, however, the addition of graphite or powdered carbon can sufficiently decrease the resistance to facilitate grounding.

The coefficient of thermal expansion of fiberglass laminate is about the same as that of aluminum, approximately twice that of steel. The low modulus, however, produces low thermal forces during a temperature change and thus allows the designer to use lighter anchors and guides when designing the support system.

The translucency of fiberglass tanks, by making the liquid level visible, is often taken advantage of to eliminate the need for gauges.

Ease of repair is an important measure of the usefulness of a material. RP can be repaired with rather simple equipment, and the use of electricity or an open flame is not necessary. This has been a significant factor in choosing RP for use in hazardous locations.

An often overlooked advantage of fiberglass equipment is its resistance to fungus and bacterial attack, rodent attack, and attack from marine organisms. The smooth inner surface of fiberglass tanks resists buildup of solids.

Summary

Of the 14 types of corrosion in metals and alloys listed in the literature, some of the more common types are the following:

1. General (where the thickness of metal is gradually reduced)
2. Galvanic
3. Aerobic
4. Pitting
5. Dezincification
6. Graphitic and intergranular

Of the various forms of corrosion, reinforced plastics are subject to oxidative attack only.

The difference between the effects of corrosive agents on metals and reinforced plastics is important to those who specify equipment. Selection of reinforced plastics, as well as other materials of construction, should be based on the installed long-term cost-performance relationship and not on material cost comparisons alone. If properly chosen for the service involved, reinforced plastics made by proper workmanship will last almost indefinitely. After a slight loss of strength during the initial period of use, reinforced plastics can be expected to remain stable and show little further deterioration of properties. Metals, on the other hand, will continue to react with the chemical environment and continue to deteriorate over a period of time.

The availability of fiberglass tanks is not a problem in North America. Qualified molders, supplying filament-wound or contact-molded tanks, are found in all regions of the U.S. and Canada. Purchasers are cautioned, however, to review the experience of respondents to widely circulated bids. It is possible that shops doing general custom work for amusement, display, or even marine markets may erroneously bid on work for applications such as those listed in Table 7-1 that require experience in handling corrosion-resistant resins and sophistication in building up interfaced laminates of different composition. Although at present there is no third-party certification system for quality assurance of equipment and/or its producers, this is a goal of the ASME project initiated by

Table 7-1. Uses of Chemical-resistant Fiberglass Tanks (partial, illustrative).

Agricultural chemical production	Pickling processes
Chemical processing (all types)	Portable processing tanks
Emergency control	Portable water
Farm storage	Pulp and paper
Filtration systems	Scrubber systems
Industrial waste	Septic service
Intermediate holding sites	Slurry containment
Ion-exchange columns	Transit tanks (road and rail)
Mineral processing	Waterwater treatment

MTI and SPI [4]. In the meantime, qualified molders may be identified through circumstantial traits such as familiarity with the standards and codes in this chapter, association memberships, and the testimonials of customers served as represented in the molder's own technical literature. A list of questions for evaluation competitive bids is found in the SPI *Users Guide* [2]. The Society also supplies lists of member producers of fiberglass tanks. For those concerned in depth with chemical-resistant materials selection, the National Association of Corrosion Engineers *Corrosion Data Survey—Nonmetals Section* [6] is indispensable.

Pressure Vessels and Tubes

Pressure vessels and tubes involve a more finely engineered category of fiberglass tanks (treated in the previous section) and fiberglass pipe (treated in the next section). Since fiberglass equipment is designed specifically for atmospheric applications, the pressure ratings required are a primary design consideration.

Normally, the equipment manufacturer is obligated to design and test the vessel and/or tube production for nominal performance at his own expense. This is not the critical evaluation. The latter occurs when the vessel of tubing has been installed, prior to connection to the operating system. At that time, all outlets must be sealed and the unit tested hydrostatically according to the following ASTM procedures, as appropriate:

D1598   Test Method for Time-to-Failure of Plastic Pipe under Constant Internal Pressure
D2143   Test method for Cyclic Pressure Strength of Reinforced, Thermosetting Plastic Pipe
D2517   Specification for Reinforced Epoxy Resin Gas Pressure Pipe and Fittings
D2924   Test Method for External Pressure Resistance of Reinforced Thermosetting Resin Pipe
D2992   Method for Obtaining Hydrostatic Design Basis for Reinforced Thermosetting Resin Pipe and Fittings
D3517   Specification for Reinforced Plastic Mortar Pressure Pipe
D3754   Specification for Reinforced Plastic Mortar Sewer and Industrial Pressure Pipe
D4163   Specificatin for Reinforced Thermosetting Resin Pressure Pipe (RTRP)

This testing should be provided for in some detail in the bid, proposal, and final contract. Since each installation is subject to variables, the contract should make clear who is the final acceptance authority, where various installation responsibilities begin and end, and which costs are part of the puchase as differentiated from late customizing requirements of the owner.

A fundamental treatment of classical stress analysis in relation to composites can be found in Chap. 13 of Lubin [7].

Pressure tubing commonly refers to systems conveying material at elevated temperatures and/or pressures. Composites are suitable for header applications and are fabricated to meet exact outside diameter and decimal tube wall thicknesses. Successful applications for composite pressure tubing include the following:

Air heaters	Condensers
Boilers	Oil stills
Headers	Superheaters
Heat exchanges	Couplings

Helically wound, portable, composite pressure vessels for liquid oxygen, gases, and transportation fuels are not within the scope of this chapter. These tanks have been fabricated integrally from composites or around aluminum liners and used successfully on high-altitude expeditions and in special vehicles. They are standard equipment in certain military and aerospace craft.

Fiberglass piping systems play a major role in chemical processing, as transmission lines for oil and gas, in municipal wastewater drainage and treatment, and in numerous industrial settings. The strength-to-weight ratio of this piping material can be varied by fiber orientation, another principal advantage of using composites. Weight considerations with respect to pipe are critical, for these translate directly into cost savings via speed and ease of installation. Still another characteristic of interest to users has been the improved hydraulics and flow factors that are achieved by the smooth interior surfaces fiberglass affords.

Three different processes are commonly used to fabricate fiberglass pipe: (1) filament winding, (2) centrifugal casting, and (3) hand lay-up. Filament winding and centrifugal casting are used to make pipe up to approximately 12 in. in diameter, with filament winding being the most common. Hand lay-up is generally used for larger diameter pipe and for asymmetric shapes.

In filament winding, continuous fiberglass filaments, called "rovings," are saturated with catalyzed liquid resin and helically wound around a polished steel mandrel. Typically, the fibers are fed through a mechanical device that moves up and down the length of the rotating mandrel. The resin is then cured at elevated temperatures and the finished pipe removed from the mandrel. Filament winding results in the highest fiber-to-resin ratio of the three fabrication methods and consequently offers the highest strength-to-weight ratio.

The centrifugal casting process involves layering glass cloth on the inside walls of a tubular mold rotated at high speed. Catalyzed liquid resin is then injected into the rotating mold. Centrifugal force ensures that the reinforcing fibers are thoroughly saturated with resin and serves to drive out air bubbles that might compromise the physical properties of the pipe. The mold continues to rotate while the resin cures. Centrifugal casting typically results in a 100-percent resin liner, which acts as an excellent chemical barrier. The resin liner also resists abrasion and protects the fiber reinforcement.

As the name suggests, hand lay-up is a manual fabrication process. It involves building up layers of chopped glass or woven glass mat impregnated with catalyzed resin around a suitable mold. Special metal rollers are used to improve glass wet-out and force out trapped air bubbles. Hand lay-up is generally used only for large-diameter pipe for which filament winding or centrifugal casting is not practical, or for custom shapes.

The variations possible for different design objectives are shown in the Table 7-2.

The same corrosion-resistant resin formulations used in other fiberglass systems are adapted to piping: isophthalic polyesters, vinylester, and epoxy. Thermoplastic linings of polyvinyl chloride (PVC), nylon, or polyethylene have been used successfully on both inner and outside surfaces to combine a specific nonstructural property with the strength of the fiberglass system. It is likely in the future that piping materials will overlap market and application more commonly. Specifiers in the next generation should not be surprised to find thermoset coatings available for cement and cast-iron soil pipe.

A harbinger of ingenuity is the rehabilitation lining system known as "in situ" (on-site and in place). In this method, a flexible tube of polyester felt fiber, impregnated with liquid resin, is cut to the inside diameter and length of

Table 7-2. Design Objectives.

Glass Reinforcement	Fabrication Processes	Approximate Percentage of Resin in Laminate
Single-strand roving	Filament winding	30
Multistrand roving	Filament winding	30
	Spray-up	70
Woven roving	Hand lay-up	60
Fabric	Hand lay-up	55
Reinforcing mat	Hand lay-up	70
Surfacing veil	Produces a reinforced resin-rich surface in many fabrication processes	90

the defective pipe section. The tubing is fed into the line at a manhole or other access point and forced through by cold water under pressure. When the inverted liner is pressed firmly against the walls of installed pipeline, the water is heated to induce cure of the resin. A rigid impermeable liner results, saving the time and expense of replacing the original system. This method has proved successful for industrial waste disposal systems as well as for municipal sewer lines of relatively small diameter.

Design criteria for piping systems comprise a vast body of technology that is continually being refined. A detailed treatment is beyond the scope of this chapter but may be found in references [8] and [9].

In certain highly-corrosive chemical processing applications, fiberglass pipe has progressed from the system of choice to designation as the only system that will provide satisfactory service. Resin suppliers continually update their charts of chemical-resistance capability, qualified by temperature and pressure. Successful installations have provided documentation for these tables and provide hundreds of entries. One authoritative third-party source is the Corrosion Data Survey (6). Surveys of general properties appear in Tables 7-3 and 7-4.

The oil and gas industries particularly appreciate the light weight of fiberglass in remote field applications. Its strength-to-weight ratio of one-fourth to one-sixteenth that of older materials means it is easier to transport and off-load. The section lengths can be longer and still be capable of being handled by manual installation teams. Fewer connecting joints per course are required, and the obstacles of terrain can be more easily bridged.

Uses of fiberglass piping uses in oil and gas transmission include flow and gathering lines, down-hole tubing and casing, surface injection piping, salt water disposal, relinings of all types, CO_2 handling, and tank batteries. The smooth interior of this pipe reduces paraffin and scale build-up. One system pulled from a depth of over 3200 feet after 23 years in service showed no degradation and was returned to use.

Because of the investment involved in chemical and petroleum processing, comparison case studies are rare, although macroeconomic comparisons are found [10]. The current pressure limitations of fiberglass pipe (roughly 3,000 psi) are cited as a limitation by oil industry users. Tensile strength is reviewed closely, and collapse strength is a particular concern. Proponents of fiberglass piping systems have a job to do in selling the service-life cost advantage of their material as compared to initial cost per running foot.

Fiberglass pipe is available in the U.S. from industrial equipment distributors in short runs of stock sizes. The number of U.S. enterprises capable of manufacturing piping systems from composites materials may exceed 100. The largest of them, the producers of filament-wound (sometimes referred to as "machine-made") piping systems, are members of the Fiberglass Pipe Institute, a trade association able to supply current names, addresses, and contact personnel

Table 7-3. Typical Thermosetting Resin Properties.*

	Uncured Resin Properties		Unfilled Castings		
	Viscosity (cps at 77°F)	Percent Styrene	Barcol Hardness	Heat Distortion Temp. (°F)	Percent Elongation
High-performance isophthalic polyester	450	50	40	165	4.2
Bisphenol A-fumaric acid polyester	475	50	38	270	2.5
Chlorendic polyester	500	30	40	284	1.4
Standard vinyl ester	500	45	35	215	5.0
High-performance vinyl ester	200	36	40	280	3.5
High-performance bisphenol vinyl ester	500	40	32	250	5.3
Standard epoxy:	13,000				
Aromatic amine-cured		—	—	320	4.4
Anyhdride-cured		—	—	312	2.5
Epoxy:	35,000				
Aromatic amine-cured	35,000	—	—	380	2.9
Furan	500	—	40	500	1.1

*From data supplied by resin manufacturers

Table 7-4. Properties of Resins Used in Fiberglass Pipe.

Resin	Chemical resistance			Other properties*		
	Acids	Bases	Solvents	Processability	Strength	Heat resistance
Polyester resins	Fair-good	Poor	Fair	Good	Fair-good	Fair-good
Isophthalic acid based	—	0	+	+	+	—
Het acid based	—	0	+	+	+	+
BPA/fumarates	+	—	—	+	+	—
Vinyl ester terminated polyesters	—	—	—	+	+	+
Vinyl ester resins	Good	Fair-good	Fair-good	Good	V. good	Good-v. good
BPA/ECH epoxy-derived						
$n = 0$	+	—	+	+	+ +	+
$n = 2$	+	+	—	+	+ +	—
Phenolic-Novolac epoxy-derived	+	—	+	—	+	+
Epoxy resins	Fair	Good	V. good	Fair	V. good	Good-v. good
Aliphatic amine cured	—	—	+	—	+	+
Aromatic amine-cured	—	+	+ +	—	+ +	+ +
Anhydride cured	—	0	+	+	+	+
Lewis acid-cured	+	+	+	—	+	+ +
Furan resins	Fair	Good	V. good	Poor	Good	V. good
Furfuryl alcohol derived	—	+	+ +	0	+	+ +

*+ + = Very Good; + = Good; — = Fair; and 0 = poor.
Het = chlorendic; BPA = Bis-phenol A; ECH = Epichlorohydrin.
Source: Based on table from M. B. Launikitis, Chemically Resistant FRP Resins, *Proceedings of 1977 Plastics Seminar*, National Association of Corrosion Engineers.

[11]. These producers quote directly to customers in their market, which they segment into three categories: chemical process industries, oil and gas production, and general industrial or civil engineering service. A much larger number of pipe producers, at least with hand lay-up capacity, is represented by the many chemical process equipment fabricators. Primarily regional in scope, these manufacturers supply fiberglass pipe and fittings as the necessary complement to their principal line of composites tanks, pumps, ducts, blower and fan, or other industrial systems. A roster of these suppliers is maintained by the resin and glass suppliers, and by the trade association as well [12].

ENERGY AND CONSERVATION INDUSTRIES

The discussion of fiberglass tanks and piping systems in the preceding sections has illustrated their applicability to energy recovery in the oil patch and gas field. This section will highlight composites applications that are valued more specifically for their energy and conservation benefits. These are numerous enough to have resulted in an international conference and volume of documentation [13].

Regardless of application, however, users of composites and those concerned with long-range planning have speculated that the use of fiberglass composites nearly always provides an energy benefit in comparison to the use of metals and concrete. By industry calculation [14], over 300 million pounds of composite laminate were used in 1984 to replace over 500 billion pounds of metal duct work, hoods, grated flooring, cement tanks, piping systems, and other corrosion-vulnerable items of industrial equipment. The energy required to produce this equipment from noncomposite materials is estimated at over 125 trillion BTUs; the equivalent consumption of energy for fiberglass materials is calculated at approximately 27 trillion Btus. This saving is equivalent to 18 million barrels of petroleum, or sufficient energy to heat 150,000 homes throughout a New England winter.

Sucker Rods

Elongated rods, traditionally fabricated of steel, are used in series (a string) and joined with metal couplings to lift oil to the surface in beam-pumping units. A sucker rod $37\frac{1}{2}$ ft long and approximately 1 in. in diameter may find itself in use on a well site as deep as 10,000 ft below the drilling surface. This application involves significant technical problems, and the string represents a major financial investment. Corrosion-resistance is the main benefit; replacement cost is reduced; downtime for service is reduced; and maintenance time is also available for increased productivity (in one Texas field, three-month maintenance schedules were extended to a two year cycle). The lighter weight—approximately one-third that of steel—reduces the stress on the string and the pumping energy required.

This application offers the opportunity to review the so-called "cascade" effect of benefits in adapting composites. The fiberglass rod's lighter weight cuts the pumping load by one-half in some cases. The producer then has the flexibility of running present equipment at reduced load, reducing the size of the pumping unit, or increasing the number of strokes per minute. In any case, the net energy needed to bring a barrel of crude to the surface is reduced. Resulting production increases will vary with conditions, but a presentation to the Society of Petroleum Engineers in 1983 documented the increase in a well operating at 5,500 feet. Barrels of fluid per day jumped from 387 to 695 when steel rods were replaced by fiberglass.

Injection Piping

The effort to go deeper and farther afield to recover oil and natural gas removes producers from the natural pressure of a petrochemical reservoir and offers a specialized challenge to the downhole tubing used for such wells. These casings must cope with the artificial "stimulants" used to drive reserves to the well head, such as gas, water injection, CO_2, and special detergents. All of these amplify the opportunity for corrosion and stress fatigue. Since 1959, fiberglass systems have proven to be the materials of user's choice in combating the combination of harsh environment and operating forces.

Geothermal Power Plants

Large-diameter fiberglass pipe has over 25 years of satisfactory use in carrying natural steam from geothermal sources close to the earth's surface [15]. This energy source, feasibly exploited in the Western U.S. and in Scandinavia, requires materials that resist the natural corrosiveness of the steam and the water generated when it cools in use. Researchers in this power plant system have found that fiberglass pipe has the added benefit of strength combined with flexibility when buried in seismically active areas.

A related potential application is the Ocean Thermal Energy Conversion (OTEC) power plant. These systems hope to generate power from the vertical temperature gradients found in the ocean. The surface water, heated by the sun, vaporizes ammonia to drive a turbine. Cold water brought up from depths then condenses the ammonia for reuse in a simple closed-cycle heat engine. The major OTEC design studies [16] suggest floating plants, but bottom-mounted installations for island supply appear to be feasible. Development of OTEC technology has slowed with the abatement of the shortage of petrochemical energy, but engineering groups view it as a viable option for the long-term future.

Temperature-Control Applications

Fiberglass industrial products have moderate insulation characteristics. The prime reason they find their way into a variety of heating and cooling applications, therefore, is their ability to handle the diverse coefficients of thermal expansion while maintaining structural integrity and resistance to the vapors, condensates, and chemical residues that accompany energy transformation. Examples of such applications are cooling towers, where flat fiberglass panels from continuous lamination lines have proven effective; heat exchanger units with elevated temperature and pressure ratings; and a variety of pumps, blowers, and air-handling equipment as well as skylights and solar paneling. It is in compression pump applications and highly engineered valve systems that the high tensile strength of graphite rather than fiberglass reinforcement finds its most prevalent use in the corrosion-resistance market.

ENVIRONMENTAL PROTECTION AND POLLUTION CONTROL INDUSTRIES

Heightened awareness of the hazards of many industrial by-products has come about during the birth of the composites industry. As the urgency of protecting the environment has provoked more action, applications have grown up alongside protection technology.

Electrostatic Precipitation

Early in the nineteenth century it was demonstrated that an electric spark would disseminate smoke in a bottle, but it took the improved electrical power supplies and control systems of the mid-twentieth century to make this phenomenon an efficient method of controlling particulate and gaseous emissions for a large variety of industrial applications. The so-called "wet method" employs a sprayed film to collect particulates from electrode surfaces and gives the process its WEP designation.

A U.S. Environmental Protection Agency report in 1976 gave major impetus to the development of composites for this control technology, as it cited corrosion problems precisely in those areas of construction manageable by composite fabrications. A second target was the reduction of maintenance. In the following decade, a wide variety of reinforcement systems and resin compounds were formulated to respond to the WEP demands of numerous acidic environments. Conductive carbon fibers, properly laminated and used with nonconductive fillers, have solved unique problems that were previously managed only with metals having a very high replacement ratio [17]. Composite WEP units have been designed for chemical incineration, acid mist collection, boiler gas pollutant collection, smelter gas cleaning, and kiln off-gas collecting [18].

Duct, Scrubber, Absorber, and Stack Liners

Mechanical and gas-fed pollution control systems also rely on fiberglass equipment. Where ceiling-suspended fume-handling is required with no sacrifice in the corrosion-resistance of a duct series, the lighter weight of composites is more than welcome. Fire-retardant systems with these same benefits have been able to handle municipal mass solid-waste incinerators with a capacity of 800 tons per day. Hydrochloric and sulfuric acid fumes reach 140° to 165°F in the ductwork of such systems. Steam hydro-scrubber units are increasingly common at chemical plants, crude-oil processing facilities, and public works. It is not uncommon for absorbers with nominal dimensions of a 6-ft diameter and 20-ft length to have as many as 30 openings for flanges, nozzles, gauges, and inspection covers. These custom features are made possible by the custom nature of fiberglass fabrication.

A power plant near Delta, Utah has the distinction of being the world's largest fiberglass application to date. Dual chimney stack liners installed within a 682-ft reinforced concrete structure use approximately 100,000 miles of fiberglass roving, for a total weight of 1.5 million pounds. This and more modest liners typically protect chimneys from scrubbing gases resulting from the removal of the sulphur dioxide produced during coal-fired power generation. Fiberglass structures of this size are often filament-wound in cylindrical sections on the plant site by mobile fabrication units.

Underground Storage Tanks

An estimated 1.4 million underground storage tanks for gasoline alone, apart from other environmentally hazardous fluids, benefit from little corrosion protection beyond their steel structure and were buried more than twenty years ago. The resulting potential for ground-water contamination has led the U.S. Environmental Protection Agency to mount a program named (with provocative acronym) Leaking Underground Storage Tanks. In some cases, fiberglass tanks built to specifications certified by Underwriters Laboratories for fuel containment can be part of the solution. There are also secondary containment systems available that employ polyester fabric extrusion coated on both sides to provide designed resistance to fuels and abrasion resistance.

AGRICULTURAL, PAPERMAKING, AND MINING INDUSTRIES

Farm Chemical Handling and Implements

Agricultural chemicals, themselves formulated with increasing sophistication, require containerization with the light weight and adaptability provided by fiberglass structures. Silos, fertilizer bins, troughs, and off-road transport tanks are among the applications. An added benefit is the translucency of fiberglass containers that allow the user to monitor present capacity. The hoppers mounted on planting equipment most often take advantage of this convenience.

Food Production/Preservation/Transport Systems

These represent growing RP applications world wide. The advantages of sanitary, cleanable surfaces, impervious to organic attack, make the use of RP attractive. Often these plastics present an aesthetic improvement as well. Extensive use, as well as laboratory extraction testing, have shown that foodstuffs are not subject to dilution or contamination from food-grade polyester-resin systems even over the long term.

A new formulation in the U.S. for ease of fabrication on vertical surfaces is thixotropic resin. It is offered for meat and poultry processing plants. In Spain, isophthalic resin systems are used to store olives during the fermentation process in solutions with a pH value of 3-12. This sodium hydroxide bath can be further temperature-controlled because the RP tanks lend themselves to underground installation.

Livestock Containment

This is a potentially large market near population centers. Fencing, pens, or flooring materials have traditionally been wooden, concrete, or galvanized steel. Aluminum offers the requisite strength and stiffness but falls short in corrosion-resistance and cost. Pultruded fiberglass slats, rods, and beams are a durable solution since they resist deterioration and are easily cleaned. Particularly where the use of holding systems is frequent and replacement is a factor, pultrusions reduce costs to a significant extent.

Pulp Mills and Papermaking

The trend to modernization and the environmental desirability of "closing in" pulping operations have motivated the development of fiberglass equipment for this significant industry. Closure practices have resulted in higher operating temperatures, lower pH levels, and increased chemical concentrations, particularly of oxidizing chlorides.

The typical pulp mill processes diagrammed in Figs. 7-1, 7-2, and 7-3 show the variety of equipment uses [19]. The shaded areas show potential fiberglass equipment applications throughout the stages of pulping, bleaching, and pollution control.

Mining

Guard rails, walkway stiles, safety ladders with protective cage, and roof support beams are among the heavy-duty uses for composites in the mining industry. Their application is suggested for (1) reduction of construction costs on difficult terrain, (2) system service life, (3) light weight, and (4) maintainability.

The industrial use of composites in mining operations suggests comment on abrasion resistance, a property not discussed earlier. Composites, with the exception of premium moldings such as the carbon-fiber-reinforced polyvinylidene fluoride used in sealless magnetic drive pumps, do not perform well in dry, abrasive applications. Attempts to convey cement, oyster shells, and even plastic bead granules pneumatically have met with early failure. The addition of a liquid medium to create a slurry, however, overcomes this limitation. Epoxy piping systems in particular, both with and without resin-rich liners, have been successful in handling foundry sand, limestone, and various gas sludges.

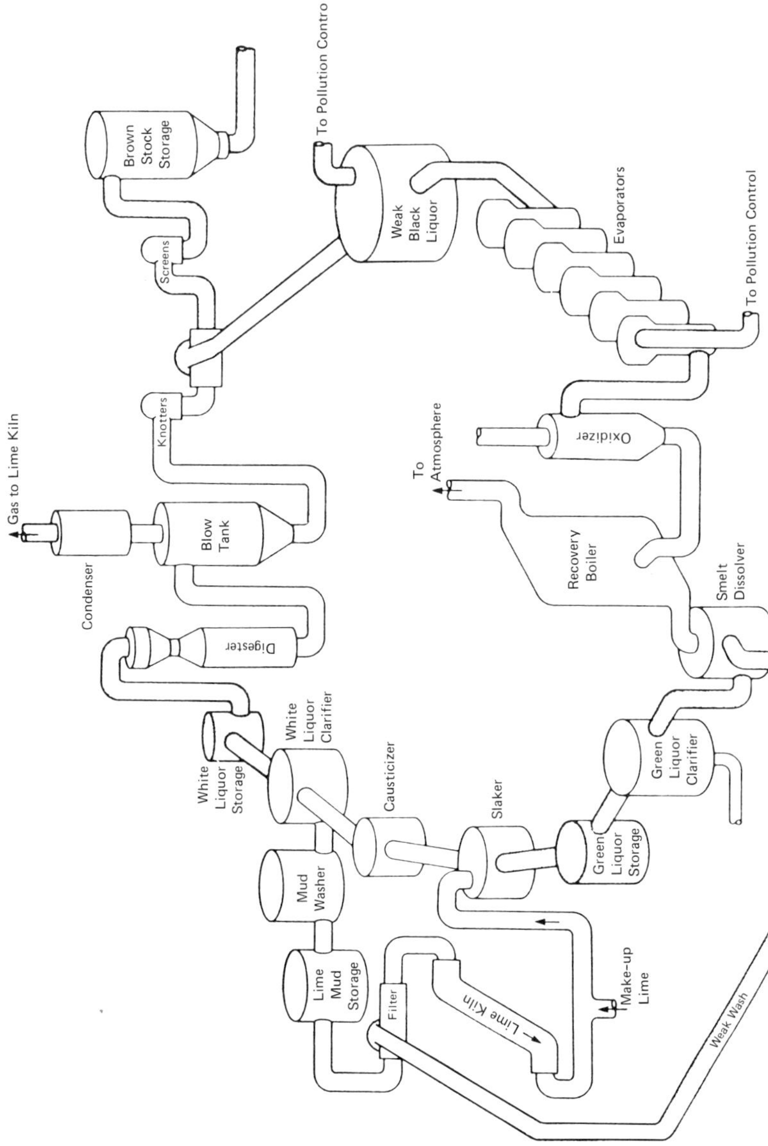

PROCESS INDUSTRY APPLICATIONS 245

Figure 7-1. Pulping stage.

246 ADVANCED THERMOSET COMPOSITES

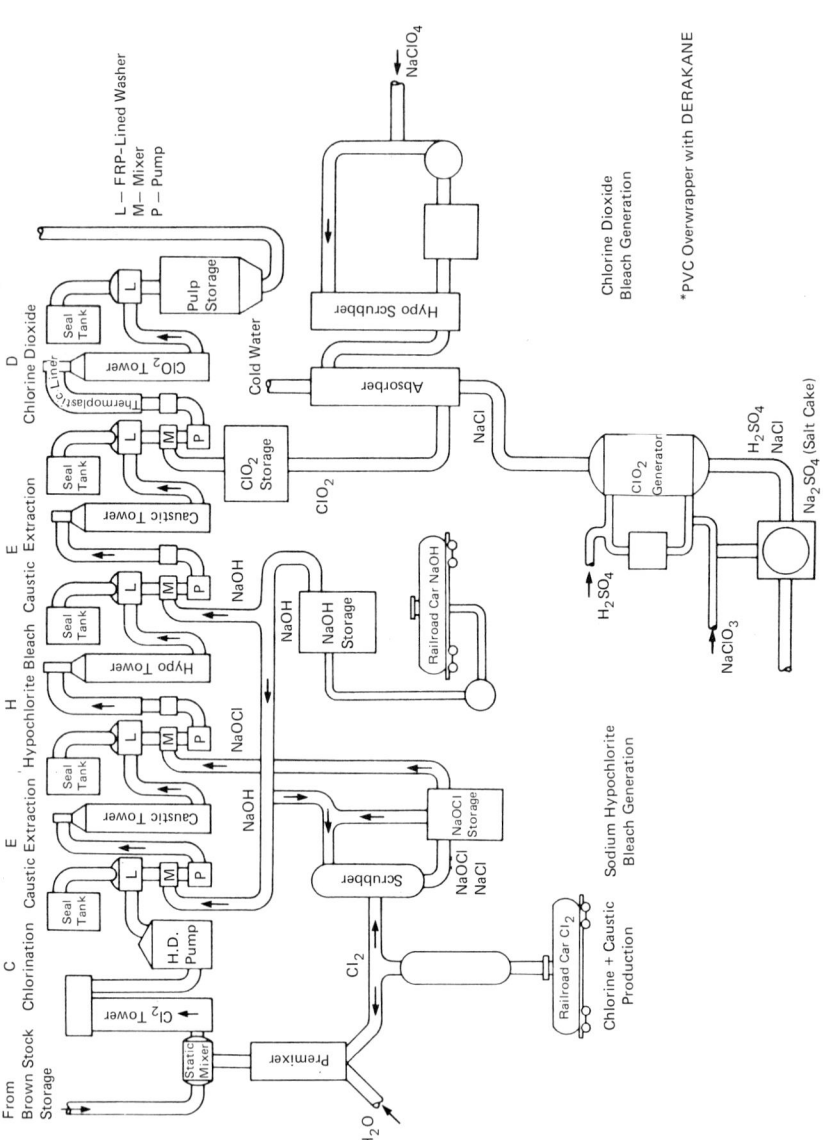

Figure 7-2. Bleaching stage.

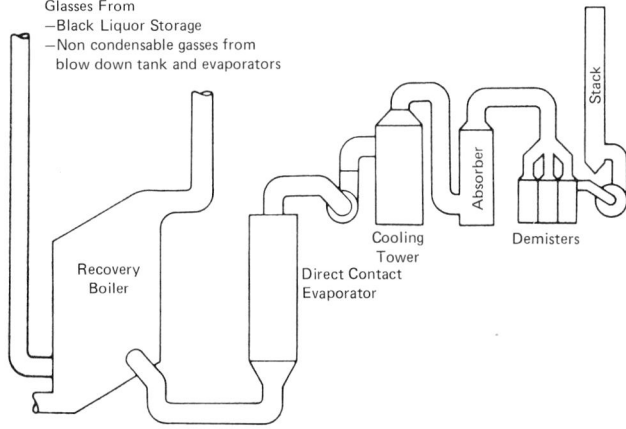

Figure 7-3. Pollution control.

CONCLUSION

Process industry applications for composites will continue to develop as technology creates new industrial conditions. The fiberglass industry's information infrastructure records installations that are over 20 years old and reports back on their viability in case history format [20, 21]. Thus, although the creativity of plant managers continues to be challenged to adapt experience to new conditions, there is a growing record of success within developed specifications.

References

1. ASTM, 1916 Race St., Philadelphia, PA 19103. Volumes of standards are reprinted annually. Fiberglass equipment standards are found in Vol. 08.04, Sect. 8, 1984.
2. Reinforced Plastics/Composites Institute, a division of the Society of the Plastics Industry, Inc. (SPI), 355 Lexington Ave., N.Y., NY 10017.
3. Cheremisinoff, N. P. and P. N. *Fiberglass Reinforced Plastics Deskbook*, Ann Arbor Science Publishers, Ann Arbor, MI, 1978.
4. Available from SPI Literature Sales, 355 Lexington Ave., N.Y. NY 10017 (132 pp., drawings and charts, glossary; $90.00 prepaid).
5. American Society of Mechanical Engineers, United Engineering Center, Hammarskjold Plaza, 345 E. 47th St., New York, NY 10017.
6. National Association of Corrosion Engineers, *Corrosion Data Survey—Nonmetals Section*, NACE, Houston, TX 77001.
7. Lubin, G. *Handbook of Fiberglass and Advanced Plastics Composites*, Van Nostrand Reinhold, New York, 1982.
8. Sessions 10 and 15, Fiberglass Pipe Institute Seminar, *Composites Go to Market*.

Preprint of the 39th Annual Conference of the Reinforced Plastics/Composites Institute, 1984, SPI, 355 Lexington Ave., New York, NY 10017. Available in Segment vol. A., $35.00 prepaid.
9. Graham, Thomas E. "Design of Underground FRP Piping Systems." 1985 Western SPI RP/CI Technical Conference Preprint, pp. 23–43. SPI, 12440 Firestone Blvd., Norwalk, CA 90650.
10. Kilkka, Kenneth. "Oil Field Pipe: Steel, and the RP Challenge." Session 18-D, Fiberglass Pipe Institute Seminar, *Composites Go to Market*.
11. Fiberglass Pipe Institute, an operating unit of SPI, 355 Lexington Avenue, New York, N.Y. 10017.
12. Corrosion-Resistant Structures Committee of SPI, 355 Lexington Avenue, New York, N.Y. 10017.
13. Institut Francais du Petrole, *Energy and Composite Materials*, June 3–5, 1981, Venice, Italy. Proceedings in English, French, German, and Italian, 502 pages.
14. "Plastics: the energy saver in industrial tanks/ducts," January, 1983, SPI, 355 Lexington Ave., N.Y., N.Y. 10017.
15. Boucher, A. and Ferre, K. "A Logical Approach to the Procurement of RTRP for Geothermal Power Plants," Pacific Gas and Electric Co., 1985 Western RP/CI Preprint, API, 12440 Firestone Blvd., Norwalk, CA 90650.
16. Hove, D. "Design and Fabrication of FRP Cold Water Pipes for OTEC Power Plants," Science Applications, Inc. 1981 Western RP/CI Preprint, SPI.
17. Jeros, Walter E. "FRP for ESP's," *Pollution Engineering*, June, 1982.
18. Sebille, J. "Modular Wet FRP Electrostatic Precipitator," Fluid-Ionic Systems, 1985 Western RP/CI Preprint, SPI.
19. Courtesy Derakane® Coatings and Resins Dept., Dow Chemical U.S.A., Midland, MI.
20. Weekly Executive Briefing, Reinforced Plastics Edition, Market Search, Inc. 2212 W. Central, P.O. Box 2886, Toledo, OH 43606.
21. See, for example, case history files published periodically by Amoco Chemicals Corp., Isothalic Resins Group; Dow Chemical U.S.A., Vinyl Ester Resins; Ashland Chemicals, Coatings and Resins Div.; Owens-Corning Fiberglas Corp., etc.

8
BUILDING CONSTRUCTION MATERIALS

Andy Green
and
Tanongsak Bisarnsin
Composite Technology, Inc.
Fort Worth, Texas

The construction industry is the second largest end user of conventional reinforced plastic composites. The majority of these products are composed of conventional composites and used in non-load-bearing applications. Tub showers and chopped strand corrugated sheet paneling constitute about 65% of these applications. The purpose of this chapter is to look at load-bearing applications of composites or advanced composites in this industry.

The main intention is to inform people in the construction industry, end users, engineers, and architects of what advanced composites are and to describe some advanced composite products available to overcome the effects of corrosion on metallic structure or conductive and magnetic materials for load-bearing or structural applications. Architects and engineers who understand the characteristics and unique properties that advanced composites can provide will significantly increase their range of material selection and greatly expand their design parameters.

A composite is defined as an assembly of dissimilar constituents intended to do a job that none of the individual materials can do alone. Chopped strand and woven roving laminates as used in shower stalls and corrugated sheeting and paneling are examples of a conventional reinforced plastic. An advanced composite is one in which the reinforcing is advantageously placed within the section to enhance its superior features and, furthermore, is installed to optimize the design and mechanical properties of the member for predictable short- and long-term performance. It can consist of a blending of various forms of reinforcing to overcome internal stress, with the overall end result of minimum deflection and high unit strength. It further consists of selecting binders that have substantial elongation (greater than the fiber reinforcement) to enhance the ability to obtain higher average stress levels. Thus we may say that an advanced composite is one which is composed of the same raw materials as its conventional composite counterpart but, by judicious placement of the reinforcing, develops substantial increase in performance. An advanced composite purlin or girt of equivalent performance to a steel purlin or girt is an example.

This chapter will only address fiberglass reinforcing in the composite because it has the highest strength and stiffness per dollar; good dimensional stability, electrical properties, and corrosion resistance; ease of handling in fabrication; and reliable mechanical properties for predictable performance of the composite. The cost of employing aramid, carbon, or boron fibers as reinforcing materials is too prohibitive for building and construction applications at this time. They are employed where high strength and stiffness-to-weight ratios are critical, such as in aerospace applications.

APPLICABILITY OF ADVANCED FIBERGLASS REINFORCED PLASTICS (FRP)

There is a strong desire to use fiberglass reinforced plastics (FRP) for permanent building and construction components, for example, cladding, load-bearing structural members, and a complete building system, because of the requirements that arise from specialty applications such as corrosive environments and microwave facilities for nonshielding or nonreflective services. In general applications, however, the industries are still reluctant to accept FRP as a permanent load-carrying material for structural components the way they accept steel, concrete, masonry, or timber. The reasons are multifold. There is still limited properties information relevant to durability and failure mechanism when FRP composites are subject to sustained load under a hostile environment at elevated temperatures, but as a long-term database for creep, stress rupture and creep modulus are accumulated so is the level of confidence in the industry. The design of FRP is further complicated by the fact that these mechanical properties are time and temperature dependent due to the viscoelastic nature of FRP. On the other hand, engineers and designers have been primarily trained in conventional elastic materials upon which the design equations, specifications, and building codes are based. The nature of price competitiveness in the construction industry puts FRP in a disadvantageous position unless their life-cycle cost is a consideration.

Outstanding performance and cost competitiveness of advanced composites for building and construction applications have been achieved by the use of unidirectional glass roving (UDR) and compatible resin system, effective shape, and proper selection of manufacturing process. Recent applications have made use of advanced composites as permanent structures which are capable of withstanding long-term permanent and repetitive loads, and weathering.

Advanced composites possess attractive characteristics, in addition to their structural properties, which enhance the overall effectiveness of the materials. At the present time, corrosion resistance is the primary reason for the selection of plastic composites. Their high strength and light weight are increasingly important when energy saving for transportable units is essential. Electromagnetic transparency is an essential characteristic for microwave-related applications. Electromagnetic noninterference is required for electronics testing facilities. Light transmission and translucency are essential features for skylights, atriums, and wastewater treatment facilities. Advanced composite materials can be formed into any desired shape for efficient stress distribution, stiffness, and other functions. They exhibit a high degree of elastic elongation and resiliency; therefore, structural parts can withstand large deformation before reaching failure. They can be repaired and restored on-site with minimum disruption of operations. These structures can be manufactured for simple assembly, minimum painting requirements, and moderate tooling cost.

MANUFACTURING OF ADVANCED COMPOSITES

The basic raw materials for the advanced composite are manufactured by large companies who are among the top 500 industrials; although in general, the manufacturers of end user products tend to be small manufacturing enterprises.

Polyesters are among the lowest cost resins available, and they are used extensively in FRP for industrial construction structural applications that require corrosion resistance at service temperatures under 160°F. Vinyl esters have superior strength and rigidity retention at elevated temperatures to 220°F, superior chemical resistance, and elongation in excess of 4% for compatibility with reinforcing glass fibers.

Additives are added to the resin to impart certain properties, but they may have detrimental effects on mechanical and other physical properties. The commonly used additives are flame retardants and fillers. Flame retardants are used to reduce burning rate but may cause problems including reduction in corrosion resistance, strength, stiffness, and UV stability. Fillers are added to improve properties such as rigidity, impact resistance, flame retardancy, and reduction of cure shrinkage, as well as to reduce unit cost.

Fabrication processes for fiber reinforced plastics depend on their function, size, quantity, rate of production, finish, and cost considerations. The commonly used processes are hand lay-up, continuous laminating, resin transfer, and pultrusion.

Hand lay-up is the simplest and most versatile fabrication process, but the production is relatively slow and the quality is dependent on skill of operators. Hand lay-up has the advantage of almost complete freedom in fiber placement and tailoring the cross section. It is also easy to fabricate very thin sections. Continuous laminating is a process suitable for high production volume of corrugated sheets of any desired length but limited maximum thickness. Resin transfer is a closed-mold process suitable for complex structural shapes that require good performance and allover finished surfaces. Pultrusion is ideal for parts that have a constant cross section such as channels, wide flanges, rods, tubular sections, and panels. Precise dimensions as well as high longitudinal strength and stiffness can be obtained, but it is difficult to tailor the section, and thin sections are limited by the physical dimensions of the reinforcement. The manufacture of a complete structure may require all of the processes mentioned here in order to achieve maximum economy and performance.

Structural Design with Advanced Composites

Advanced composite materials are nonhomogeneous, anisotropic, viscoelastic, nonductile, and temperature sensitive; they have varying degrees of chemical resistance and are subject to weathering. They are fabricated from a variety of possible compositions and numerous processes by fabricators of varying degrees of skill. Consequently, the designer has the responsibility of analyzing the effects of stress situations that go beyond the basic static strength and rigidity calculations required of conventional materials like steel and concrete.

These conditions (an infinite variety of material configurations) have created a situation that limits the amount of hard data available to make a comprehensive structural analysis. However, there are published data and design guides available that can provide adequate information to conduct a preliminary stress analysis and then test within the practical considerations imposed by time, cost, and risk factor (life and hazard) to confirm the design.

Design Considerations and Specifications

The advanced composite for long-term load-bearing application requires structural integrity and durability. The inner layers of reinforcing fibers are designed according to load and service requirements, and satisfy the requirement of structural integrity. The exposed layers are either UV protective coats or resin-rich layer for corrosion and weathering resistance, and are there to satisfy the durability requirement.

Factors which should be considered when evaluating a composite and used to prepare the specification for load-bearing structure include identifying the functions of the component structurally and other requirements, such as chemical, thermal, electrical, and aesthetic characteristics. Then, the following factors should be evaluated:

Nature of loading. The strength and rigidity properties of FRP are dependent upon whether the applied load is intermittent, cyclic, impact, or sustained. Mode of failure, which varies directly with the ultimate and buckling strengths, is associated closely with the type and duration of load.

Environmental Conditions. Service environments, such as UV radiation, weathering, aggressive chemicals, and elevated temperature, have deleterious effects on the performance of the designed structure. The relevant properties are to be reduced accordingly.

Expected Life. The expected life of the component is an important parameter in determining the design values because of the time dependent nature of plastics.

Factor of Safety. Partial factor of safety for materials is applied to reduce the expected values of strength and rigidity which may arise due to variation of materials and fabrication. Partial factor of safety for loads (or load factor) is applied to the probable increases in load over the expected values based on the degree of certainty in load estimation.

Serviceability State. Service conditions must be considered in the design of advanced composites because they are the limiting conditions which are expected during the operating period of the structure. They include maximum deformation, based either on a fraction of span or on definite limiting values; nondamaging buckling and wrinkling of thin parts of the structure resulting from temporary or prolonged stresses; crazing, microcracking, and weeping, which may indicate possible deterioration in performance of the part.

Design of FRP Structural Components

Preparation of specifications is the first step in the design procedure. Specifications of the part are written to establish the standards for acceptable levels of performance and quality control.

The design of parts and structures using advanced composites must employ the proper constituents placed in the most advantageous manner. Optimum performance is obtained by the proper combination of four critical factors: shape, reinforcement, resin, and process.

SHAPE	REINFORCEMENT
PROCESS	RESIN

Shape: First size the member to support the loads, then design the cross section to maintain its shape. This requires placing the reinforcing elements (the primary load-carrying constituent) in a leveraged position to take advantage of its superior strength and stiffness.

The shape must be designed not only to satisfy the geometrical requirements of the function it performs but also provide inherent stiffness and strength. For example, a box structure is inherently more stable than a wide flange or I beam. Tubular shapes are recommended because these shapes provide optimum area to inertia properties by concentration of material along the perimeter, thus giving the maximum radius of gyration for a given cross-sectional area.

Reinforcement: The reinforcement system must be chosen to provide maximum thickness and strength as far away from the neutral axis as possible to gain advantage. Reinforcement can take several forms: woven, unidirectional, knit, combinations, or chopped fibers. The location of the reinforcement and its percentage at those locations need to be determined. This step is most important as it is imperative to determine the proper fiber count in order to insure predictable strength and stiffness. This is comparable to the steel reinforcement bar call out in reinforced concrete structures.

Resin: The resin used can have special requirements, such as fire retardance, chemical resistance, and/or UV resistance. The qualities to be imparted to the composite are determined by the use for which the end product is intended. Resin-rich surfaces and veils are used to increase durability.

Process: Once the three preceding factors have been decided upon, the manufacturing process to be used may be considered. The process will depend on the configuration of the piece and the production quantities needed.

These four factors must be carefully considered to obtain the optimum performance of the product. However, it may be difficult to do this if you do not know the types and thicknesses of the reinforcing materials available, the type of forms that may be used, or whether an acceptable manufacturing process exists. Early consultation with a reliable manufacturer should enable you to design a product that will meet your specifications and give you the optimal cost. Finding a company familiar with these materials will help you to use them to the best advantage.

CONSTRUCTION APPLICATIONS

The building and construction industry has used advanced composite materials as load-bearing structures for complete buildings and building components, complete cooling towers, cooling tower components, personnel bridges, and ancillary items. The structures and components described here were in competition with conventional materials. Award of contracts was made on the basis of "evaluated bid", which is the standard method for the construction industry.

BUILDING CONSTRUCTION MATERIALS

Figure 8-1. Restoration project: Severely deteriorated facility (on left) is replaced by advanced composite cladding, advanced composite purlin and girt structural members supported on 24 foot centers.

Low costs were available because of the effective use of raw materials by optimum design of the components.

Contractors and builders who were at first reluctant to use these materials have become supporters after completing their first job. Their claims were cost savings because of faster erection and lower labor costs, elimination of heavy equipment and the associated cost of operators, and use of lightweight and easy to handle parts. Fewer parts meant fewer joints and fasteners required and, finally, a smaller number of damaged and scrapped parts.

Cooling Towers

Cooling towers are required to resist wind loads, earthquake tremors, and load from the tower fill under corrosion induced conditions at elevated temperatures. This system consists of structural walls for cladding and supporting the fan, fan deck and stack, and beams to support the ceramic fill. The wall system has internal columns which transfer the total load to the concrete foundation.

256 ADVANCED THERMOSET COMPOSITES

Figure 8-2. Advanced composite cooling towers on high rise building. All components are of advanced composites—stressed skin, columns, and beams.

The beams of the cooling tower demonstrate the long-term integrity of advanced composite structural components under sustained load. The beams, after years of operation, exhibit no visible settlement or signs of deterioration.

Advanced composite cooling towers offer lightweight modular units which are convenient to transport and assemble. They are maintenance free, corrosion resistant, and fire retardant.

The towers are manufactured by hand lay-up processes and pultrusion.

Requirements:

1. Light weight to reduce dead load on building super-structure
2. Ultraviolet ray (UV) resistance
3. Good weathering characteristics
4. Unit construction or pre-engineered for fast and easy erection
5. Ability to support 57,000 pounds of tile loads (plus any intermittent ice buildup)
6. Resistance to 30 pound per square foot (psf) wind loads (about 100 mph)
7. Earthquake resistance (resist seismic loads)
8. Resistance to corrosive elements in warm water coolant (about 110°F)
9. Factory Mutual fire rating

Solution:

1. Ribbed wall panels are stressed skin with woven roving reinforcing. The columns are integrally molded with UDR as primary structure.

2. Tubular section beams provide resistance to lateral bending. There is a high count of unidirectional rovings on the flange to resist bending loads.
3. All surfaces have a gel coat and a resin-rich mat surface to resist UV and water penetration.
4. Brominated polyester resin is used for fire retardancy.

Walkway Bridges

Advanced composite walkway bridges are used at wastewater treatment facilities in chemical plants. They are used by personnel for access to aeration equipment at the center of the pond. Their steel counterparts, because of the highly corrosive atmosphere of these plants, require frequent inspection, sand blasting, painting, and replacement in three to six years.

Advanced composite bridges weigh one-third as much as their steel counterparts and can be prefabricated in one piece by hand lay-up. The FRP bridge can span 90 feet without any visible settlement during its service period. This is possible because of the selection of structural sections, and continuous reinforcing fibers which act similarly to steel rebars in concrete structures.

Figure 8-3. Seventy foot clear span personnel bridges over wastewater treatment facility.

Requirements:

1. Ability to sustain a 75 to 125 psf live load and 20 psf dead load
2. UV resistance
3. Structural stability (compression rail, web buckling, and flange crippling)
4. Resistance to rolling stability of entire bridge
5. Length/deflection $(L/D) = 360$
6. Minimum 15-year service life

Solution:

1. Stressed skin construction with a torque box is employed to resist rolling instability.
2. Gel coat is used to protect from UV and corrosive chemicals.
3. Hand rail serves double duty as load-carrying member as well as safety rail.

Pipes for Cooling Tower Water Distribution System

Circulating water in cooling towers is pumped to the top of the tower and sprayed over ceramic tiles to remove heat. These pipes must be able to resist the corrosive elements in water; however, they do not have to withstand high pressures. Holes must be cut in the bottom side of the pipe to permit the attachment of spray nozzles. The top of the pipe is flat to support the mist eliminators.

A unique cross-section design allows the pipe to be made of advanced composites, keeping weight to a minimum while meeting all of the other requirements.

This pipe is a typical example of the considerations made and used in placement of reinforcement. The two sections weigh approximately the same, but longitudinal reinforcement in the 8 inch diameter pipe is two times that of the 12 inch, while circumferential reinforcing in the 8 inch diameter is only one-third that of the 12 inch.

Requirements:

1. Low pressure (maximum pressure is less than 10 foot head)
2. Loaded in bending from mist eliminators and water at 100 lb/lineal ft (plf) on 11.2 foot span.

Figure 8-4. Photo of tower from inside.

3. Holes in bottom of pipe to install spray nozzles
4. Operating temperature 32°F to 120°F
5. Short-term deflection limited to $L/D = 600$
6. Sizes 8 and 12 inches in diameter
7. Long-term deflection limited to $L/D = 300$
8. Fire retardance

Solution:

1. Add integral high strength internal ribs to help react bending moment and to minimize the effect of stress concentration and reduction of the net section.
2. Use circumferential fibers to maintain shape of cylinder.

260 ADVANCED THERMOSET COMPOSITES

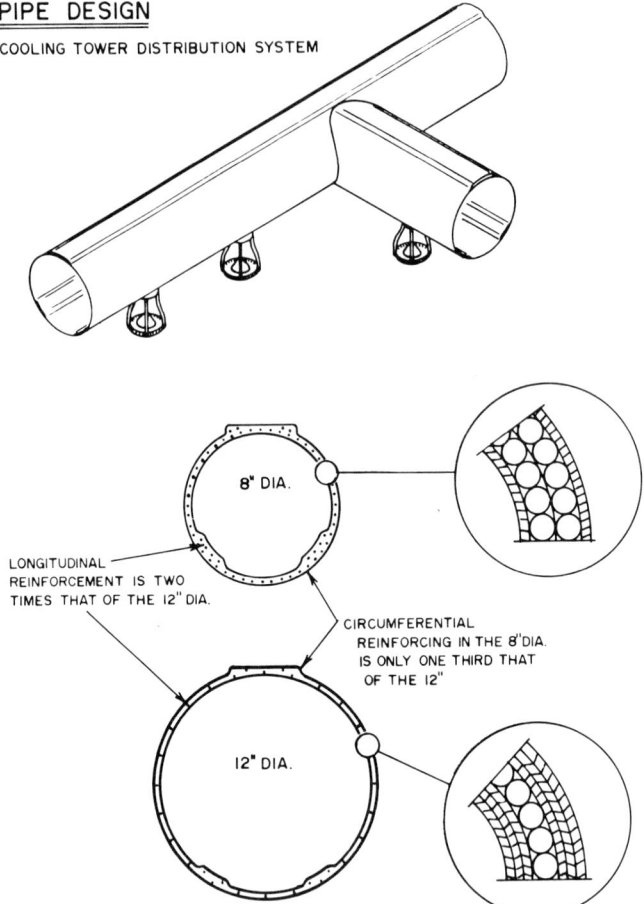

Figure 8-5. Pipe design — cooling tower distribution center.

3. Use veils on outside surfaces and inside surfaces.
4. Use pultrusion process.

Beams

Composite beams are used to support panels and decks in industrial buildings exposed to chemical environments. They are designed to replace conventional steel shapes. To overcome the low modulus, composite beams are deeper than their steel counterparts but taper on the ends to maintain a typical building profile.

The box section is laterally stable to overcome lateral stability limitations of

Figure 8-6. Composite structural roof beams, ridge vents and walkways for use in permanent facilities located in corrosive atmospheres, are light-weight, easy to install, easy to maintain, and competitive with conventional materials. The beam uses a "torque box" shape for resistance to lateral bending.

flanged structural sections. Continuous glass rovings are employed as the major load-carrying element in the way steel rebars are used in reinforced concrete beams. The results are advanced composite beams with the load-carrying capacity of rolled steel sections that are more than twice as heavy.

These beams are primarily bending elements comparable to bar joists but without the lateral buckling problem and bracing requirements.

These beams can be manufactured by either hand lay-up or pultrusion process.

Requirements:

1. Ability to be competitive with coated steel rolled wide flange
2. Ability to maintain building profile of existing usable space
3. Resistance to caustic chlorine atmosphere
4. Fire retardance
5. Resistance to structural stability failures
6. Span of 20 to 25 feet with L/D of 360 to 240 at 200 plf

Figure 8-7. A. Detail of purlin and girt at end of industrial building. B. Detail of purlin and girt at rigid frame.

BUILDING CONSTRUCTION MATERIALS   263

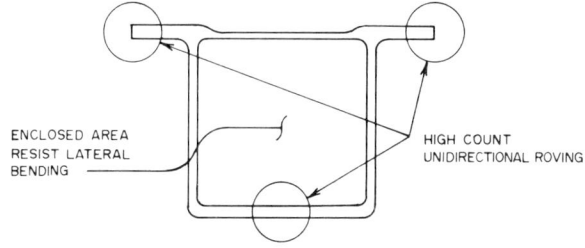

BEAM DESIGN

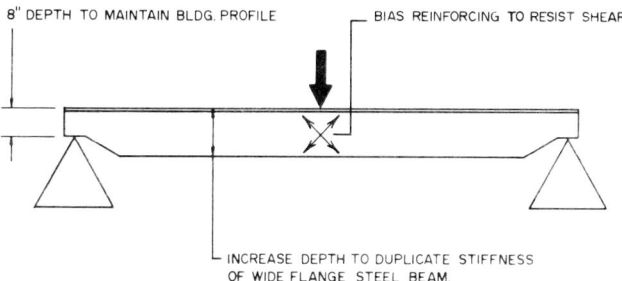

Figure 8-8. Beam design.

Solution:

1. Hand lay-up or pultrusion and fabrication
2. Closed stressed skin cross section
3. Tapering at supports
4. High concentration of UDR at top and bottom

Panels for Building Cladding

Profiled structural panels have been used to replace metal or cement asbestos siding and roofing in the highly corrosive environments of numerous industrial plants, such as cogeneration, mining operations, and chemical process plants.

Requirements:

1. Ability to sustain positive live load to 60 psf on spans to 7 feet
2. Fire retardance
3. UV resistance

Figure 8-9. These photos illustrate superiority of advanced composites for erection of long panels. Panels are lightweight, yet large elastic elongations prevent crippling of corrugations during erection and give the end user the benefit of no joints in each panel.

Figure 8-9. (Continued)

Figure 8-10. All advanced composite curved roof and ridge vents.

4. Chemical resistance
5. Length/deflection = 90; high enough to prevent fastener hole enlargement caused by large deflection—thereby matching performance of metal panel and reducing opportunities for leaks

Solution:

1. Continuous laminating process with unidirectional fibers
2. Transverse fiber and chopped strand
3. Fire retardant resin
4. Both polyester and vinyl ester resins

Decking

Roof decks employing continuous glass reinforcement are used as a structural part of a built-up roof system for sustained dead load and live load. The decks are designed and manufactured to meet the same criteria used for metal decks in regard to deflection limits and strengths.

The applications of advanced composite decks are similar to those of panels. They have been used in chemical bulk storage and caustic chlorine facilities.

(a)

(b)

Figure 8-11. (a) Installation of advanced composite decking. (b) View from inside.

Both panels and decks are manufactured by a continuous laminating process; heavy decks are made by pultrusion.

Requirements:

1. Ability to sustain a dead load of 20 psf (long term) and live load of 40 psf
2. Ability to sustain a 90 psf uplift load
3. Fire retardance
4. Chemical resistance
5. Minimum span of 7 feet, maximum of 10 feet
6. Low deflection: $L/D = 240$
7. Heat resistance to 150°F

Solution:

1. Continuous laminating and pultrusion
2. Veil protective surface
3. High concentration of UDR on flanges
4. Transverse fibers and mat in webs
5. Vinyl ester resin

Pre-engineered Buildings

Industrial buildings used as testing facilities for computer and electronics products are required to be made of nonmetal materials for electromagnetic interference-free environments. Structural components are designed as stressed skin structures using torsion box configurations as in aircraft design. Sidesway deflections are kept within limits by using fixed end supports.

Pre-engineered buildings are fabricated using hand lay-up rigid frame bents and the components described earlier, then assembled with composite mechanical fasteners and bonding.

Requirements:

1. Meet local building code
2. Be fire retardant
3. Be electrically nonreflective or conductive
4. Meet pre-engineered metal building requirements for structural integrity

BUILDING CONSTRUCTION MATERIALS 269

Figure 8-12. Pre-engineered buildings will help architects, engineers, and designers because they will no longer need to go through the detail design phase normally required in both construction and plastic design.

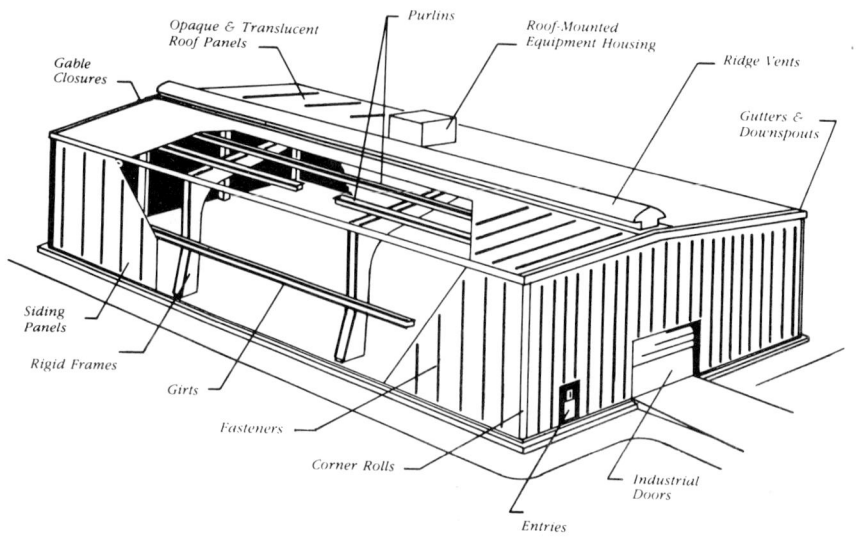

Figure 8-12. (*Continued*)

Solution:

1. Fire retardant vinyl ester resins for long-term performance
2. Rigid frame fixed at foundations
3. Rigid frame hand lay-up closed section with heavy concentration of UDR at top and bottom
4. Fabricated sections from stock purlins, girts, and panel shapes

OUTLOOK

These applications of advanced composites for a niche market were successful because they provided lower life-cycle costs or offered a service unavailable with conventional materials. The potential for growth should be good because the market has been here for many years but has not been serviced. Taking an approach that utilizes the unique characteristics of the material, rather than forcing the material into patterns derived directly from conventional structural building materials, will increase the market because end user cost will be reduced.

The future for expanded markets in load-bearing applications will depend on the dissemination of more data on performance and availability to end users, engineers, architects, and contractors; and on the availability of higher modulus reinforcing at costs competitive with glass fibers. In the case of glass fibers, the practical specific stiffness to cost, is rated two times that of light gauge steel and four times that of rolled steel. Ranges down to 1.25 will be required to make significant inroads into other markets. Regardless of the availability of lower cost and high modulus reinforcement, the industry is growing and will continue to grow because more and more is being learned about these materials and their performance in load-bearing structures for specialty applications.

9
OPPORTUNITIES FOR DESIGN ENGINEERS

Robert E. Coulehan
President
OMO Corporation
Dobbs Ferry, New York

INTRODUCTION

Whether the object is an airplane, an automobile, a bridge, a satellite or a myriad of other structures, it is the design engineer who takes the customer specifications and reduces them to practice. He specifies the materials, sets material standards, and certifies the performance of the product. His tools include design concepts, computer aided design programs, material properties, and test facilities for both materials and structures. Of course, there is the basic education and apprenticeship that each engineer will have in order to perform his job. The driving force in the fulfillment of the job is the desire to push the performance frontiers of structures to new areas of efficiency and cost effectiveness. At times, this can be a difficult learning experience for both the design engineer and the customer.

A CASE STUDY

The first regularly scheduled commercial jet passenger service began in 1952 when the British put the De Havilland Comet into operation between London and Johannesburg, South Africa. It was hailed as another great achievement in man's continual quest to fly. The military had flown jets with great success for almost ten years, and now it was time to bring them into commercial aviation. There had been major advances in the jet engine in terms of thrust and efficiency. However, this was not the case for aircraft structural materials which still reflected the design parameters developed in the thirties.

The Comet suffered a series of disastrous crashes that mystified investigators. Following the crash of a Comet off the coast of Italy, a major portion of the fuselage was recovered. An analysis of the structure showed that some sections had experienced metal fatigue due to continual pressurization cycling. Out of this tragedy, the designer became acutely aware of the need to improved design concepts, material parameters and characterization that would allow him to design for extreme environments.

That was over 30 years ago, and more advances have been made in the design of structures with all types of materials then had been made in the previous 300 years. The design engineer now has access to many more materials and their properties then this predecessors had. He also demands more information about the materials he uses. His job requires that he coordinate the efforts of all involved from the design through the fabrication and evaluation of the structures. It is his responsibility to keep up with new developments in materials, sophisticated analyses of the behavior of structures under high stresses, cyclic fatigue and many other factors that affect the design of the product.

In the example of the De Havilland Comet crashes, it was metal fatigue that was determined to be the cause. Once that was known, the remedy was easy. When a crack develops in a metal, failure is predictable. The fuselage structure of the Comet was modified to eliminate the development of fatigue cracks. As long as metals are used, the design engineer has ways of eliminating the crack propagation problem. This is not the case with advanced composites. By their nature, they are complex and have voids that can only be reduced to a minimum. It is almost impossible to predict failure of a composite based on the observation of cracks and voids alone. There is a considerable amount of work in the area of failure criteria, and some sophisticated analyses are used to help predict failure limits. From these analyses, the design engineer sets the design allowables and puts in motion the building of a new product.

Despite the problems associated with advanced composites, many new uses are found each year and many more are proposed that challenge the performance of these materials. It is through such challenges on both sides-the structural design and the materials properties-that the design engineer orchestrates the ever changing world of structural products.

Where will the new challenges arise in the next 30 years? They will come from the successes of today's designs. Each success points out how materials can be used and what changes are needed to push the materials performance to new and higher levels. This book highlights those areas in which the design engineer will serve as the director to create new structures from different composite materials that will lead to new levels of material performance.

ELECTRICAL AND ELECTRONICS

Electrical and electronic applications, and related industries such as telecommunications and computers provide continuing opportunities to design engineers. These industries offer new challenges to discover more advanced composites as shown in Chapter 4, "Electrical and Electronic Applications."

The effort to package more data processing capacity into less space has by no means tapered off. This effort toward further miniaturization could dramatically alter conventional microcircuitry as it is recognized in the mid-1980s. Innovative circuit designs of semiconductor devices is motivated by the invasion of electronics into mobile units such as transportation (see page 174) and aerospace vehicles.

New concepts can be expected to develop from current uses of high performance advanced composites and ceramics, but material selections in the 1990s could be dramatically different from today's materials.

Modern printed circuit boards (printed wiring boards) are composed primarily of epoxy resin matrixes and fiberglass mat reinforcements, and copper plating. The packaging heirarchy consists of either two single-layer composites or multilayer composites. When electronic functions exceed the normal capacity that can be packaged on one panel, two panels may be joined by connectors and cables as noted on page 120. The material selections and connector designs depend on the functional and environmental requirements of the device. These material and design considerations are far from established. Opportunities now are looking for imaginative designers to solve the requirements of printed circuit board/connector/cable systems.

Another area of opportunity for design engineers is the alternative approach to eliminate the component package concept and produce a multi-chip component package, as noted on page 120. Here, too, are exciting opportunities for interdisciplinary design engineers.

The requirements that are developing for electronic applications vary from miniaturized television sets to high-temperature aerospace applications. The automotive industry is the largest single user of custom designed integrated circuits as noted on page 174.

Onboard computers under-the-hood in passenger cars in the late 1970s and early 1980s were primarily limited to emission control and fuel economy. The rapid increase in automotive microprocessors under-the-hood; in the dashboard; and in-situ for braking, steering, and suspension all create new design requirements.

These widely varying requirements for electronic designs create a need for new epoxy and polyimide grades of advanced thermoset composites. The use of high temperature and flexible printed wiring boards extend the designer's search for new materials to high performance engineering thermoplastics.

Materials and design requirements provide tremendous opportunities to engineers who are looking for new challenges!

TRANSPORTATION

Advanced composites have been used successfully in a wide variety of vehicles. From sophisticated aircraft to trucks and cars, and to pleasure boats, all have benefited from the properties of composites. The aerospace community has been the most aggressive in setting up programs that would lead to advanced composites in many applications of advanced aircraft structures. Many components have been fabricated, tested, and flown successfully as part of the effort to promote these materials in areas where weight savings will be significant. These successes have been due to the design engineers' efforts to set design limitations and the evaluation programs to ensure performance and safety.

While there are a few primary structures that have use advanced composites, these have all been in small components where the material is well characterized and designed with a significant safety factor. An example is the tail rotor blade on many helicopters. The real challenge for the design engineer is to design a primary structure in a critical load-bearing area such as the wing of an aircraft. Now the advanced composite community has an objective that it has not faced since these materials were first introduced. All the design parameters have been established and the test programs outlined for these applications. This will not be a project that will have only one success, but rather a number of successes with a variety of composite materials.

An area in which composite materials were successful from the earliest applications was in marine uses, particularly pleasure crafts. These materials were very successful in the corrosion environment of the sea. Currently, with new resin systems and inproved fibers, the design engineer can now set up design configurations to take advantage of the improved properties for more efficient hull and structural configurations. Advanced composites above the waterline help reduce weight to keep the center of gravity as low as possible. Much of the traditional equipment currently in use can be made more effective with advanced composites. Such items as panels and storage containers made using sandwich panel construction offers weight reduction and, at the same time, high strength and stiffness. Again, the design engineer has a variety of materials that permit a greater flexibility in design than was previously available with metals and wood.

The automobile industry continues to look for advanced composites that are cost effective. As with marine uses, corrosion is a major consideration in the design of components for automobiles. The problem with advanced composites has been the time-consuming processing methods needed to obtain a finished part. It was not possible to fabricate the large volume of parts required by the industry. Hence, their use was limited to custom models. However, several fundamental changes in the automotive industry offers exceptional opportunities for advanced composites. In the past Detroit has relied on the high-volume car as its basic product. This meant that the expensive stamping mills could be paid back from this high-volume production. This is no longer the trend today. The marketing of cars emphasizes the limited production, higher mark up pricing to remain competetive in the world markets. Now, with production runs of 200,000 to 400,000 units, it is possible for the advanced composite materials to be very competetive with the traditional metal car. One example of a part that has changed the thinking is the composite leaf spring. For light and medium trucks, the metal leaf spring was the most serviced part. Now with the introduction of the composite leaf spring, this will replace the metal leaf spring within 5 to 10 years. The designer now has some successes that can be used to integrate more composite materials into the structure ofthe automobile.

SPACE STRUCTURES

One of the most challenging areas for the design engineer is the development of large structures for space applications. The space environment places special requirements on both the design and the materials used in construction. Currently, there is a large data bank of information on such structures as satellites, antennas, space telescopes and the space shuttle program. However, the large space structure of the future will be constructed in space and will place even greater demands on the use of materials and on design concepts.

In the design of space structures, three criteria must be satisfied: overall design, structural design, and operational design. Each of these design areas overlaps and must be solved simultaneously. Advanced composite materials will play an important part in the design and construction of these structures. Special fabrication and packaging techniques must be developed that will reduce the contamination to outgassing of assembled systems

The assembly of structures in space is an emerging technology that will bring a new era of space exploration during the next generation. The design engineer will play a key role in the success of these structures. The use of advanced composite materials will permit structures with high rigidity and low density that cannot be met by other materials. Good vibration damping must be designed into the structures to overcome problems of setting up and changing orbits.

Dimensional changes resulting from large temperature fluctuations is an area requiring special design parameters. As a space structures rotates or pass into and out of the shadow of the earth, dimensional changes can occur due to large temperature variations. Some progress has been made in this area. In the fabrication of the space telescope that is scheduled to be launched in 1986, a special design was used in the tubular structure that took advantage of the negative coefficient of expansion of graphite to keep changes in the focal length of the structure to zero during temperature fluctuations. In larger space structures, new designs will be needed for more complex structures that will be serviceable over extended temperature cycles in a space environment.

Radiation exposure will be a major problem with the use of advanced composite materials in a space environment. These materials must have some protection against cosmic and UV radiation of much greater intensity than found on earth. There are data available on selected composite systems that can be used as guides to predict behavior in space. Such data will have to be updated to include many of the newer systems. In addition, as space structures are sent into higher orbits, some radiation protection will be needed for the crew and space workers. Journeys on one to five years will require structures that protect the life support systems with high reliability.

A major area for the design engineer will be the use of deployable and erectable structures in space. The design engineer must create new ways of packaging the components for shipment and assembling the structures in a weightless environment. Features such as light weight, storability and the ability to survive the launch profile acceleration will be principal parts of the specifications for such structures. As part of the packaging design, components such as tubes and boxes and act as containers for other parts. All these tasks are the domain of the design engineer

Designing an electrically charged free structure is most important when using advanced composites. There are techniques available today for these systems that are used in aircraft exteriors. In larger space structures, this problem will be magnified and require new ways to conduct charges away so as not to have, in effect, a large capacitor.

Though actual work on large space structures is not in progress, some studies have been carried out as a preliminary to determine some of the problems that must be solved. The design engineer will have a continuing role to play throughout the development of such structures so as to meet the objectives of the program.

ROBOTICS

The emergence of robots begin a new industrial era in the world. Great strides have taken place in this field that will continue to create new opportunities for the design engineer. It will be up the design engineer to make these units more efficient and trouble free for extended periods of operation. There is a great opportunity to use advanced composites to make the robots more cost effective.

While much emphasis has been placed on robots for carrying out routine operations, an area of equal importance involves work stations in harsh environments. While conditions such as contaminants, corrosive solvents and high temperatures or humidities make many operations hazardous to humans, robots could carry out the work with greater efficiency and at a lower cost. The design of robots in which filament wound and laminated components are used to provide both impact resistance and corrosion resistance will be paramount. It is the design engineer who will create the design for the most cost-effective unit. Taking advantage of the new materials and processing techniques will enhance the operation of the robot and make it more acceptable in many applications.

Packaging of the robot's electronics, computer and power electrical equipment must be done with a great deal of sophistication. Keeping the low power systems operational at all times will be a priority, for it is from these systems that all commands will be issued. Some type of laser will most likely be included in many of the designs. In addition, the design must include different levels of safety to ensure that the robot's operation will remain in a fail-safe mode.

INDUSTRIAL PROCESSING AND CONSTRUCTION

The broad industrial markets require storage and transportation of liquids for processing and, at the same time, need to be cost effective. Advanced composite materials can play an important role in defining new products for many applications. Many pipes and storage tanks currently use composite materials but are restricted from certain applications because of high temperature and pressure requirements. Part of the problem lies in the design and fabrication methods currently in use. New resin systems are now available which, when properly processed, can operate at higher temperatures and pressures for extended periods of time. In such applications, it is important for the design engineer to ensure that the design configuration is reduced to practice without compromising the structure. For a pipe system, through proper design it is possible to create a cost-efficient operational product that meets all the design specifications. For such industries as petroleum, paper, textile, mining, agricultural, chemical processing, and others, designing for a particular environment using selected materials and design concepts will be of major importance to these industries. All designs for the processing industry require conversion of materials to a new form where energy is an essential part of the process. It is here that advanced composite materials can be most effective if the design is to be competitive. Many other areas in the industrial market will be able to take advantage of the results of using advanced composite materials in the traditional pipe and tank configurations.

CONCLUSION

Today, the design engineer has been give a wide range of responsibilities because of the increased number of materials that exist. The choices that are made must be based on how effective the material makes the design. To ensure the performance of the product, the way in which materials are processed now become the domain of the design engineer. In each chapter of this book, this role has been emphasized and made specific for a particular area.

INDEX

Aircraft/aerospace applications, 2, 7, 9, 16, 20, 22, 24, 25, 27, 31-33, 38, 39, 41, 42, 47, 58-60, 62, 65, 66, 68, 69, 74, 83, 84, 94, 142, 143, 177, 179-182, 193-226, 235, 272, 273, 275-277

Aramid fibers (Kevlar fibers), 2, 5, 31-34, 54, 68, 75-77, 83-95, 98, 105, 107, 126, 166, 177, 193, 197, 199, 201, 210-213, 216, 217, 220, 227, 250

Automation, 38-41, 47, 50, 56, 64-66, 68-70, 181, 182, 226

Automotive applications, 9, 12, 24, 33, 38, 39, 56, 58, 64, 65, 69, 83, 111, 163, 164, 174-192, 275, 276

Boron fibers and filaments, 1, 2, 5, 25-28, 33, 54, 68, 105, 106, 198, 250

Building and construction products, 9, 12, 38, 56, 69, 249-271

Bulk molding compound (BMC), 12, 62, 64, 79, 80

Carbon graphite fibers, 1, 2, 5, 6, 9, 10, 17-19, 25, 27-32, 34, 49, 54, 58, 59, 62, 68, 75, 83, 87, 94-105, 166, 177, 181, 183, 185, 193, 194, 197-201, 203, 205, 207-217, 219, 220, 227, 242, 244, 250

Casting process, 12, 235, 236, 238

Compression molding (see also Sheet molding compound, SMC), 8, 79, 176, 178, 186, 188, 189, 200

Contact molding, 228, 230, 233

Coupling agents, 23, 134-138, 140, 141, 144

Curing, 3-9, 11-13, 15, 17-20, 23, 38, 49, 52, 53, 60-62, 65, 79, 138-142, 161, 191, 195, 197, 203, 204, 209, 210, 213, 217, 218, 229, 236

Electrical/electronic applications, 20, 24, 49, 56, 69, 83, 110-167, 274, 278

Epoxy resins, 1, 3-7, 11, 13-15, 19, 20, 24, 26, 27, 30, 31, 33, 34, 58, 74, 79, 82, 83, 85, 92-94, 102, 104, 110, 125, 127, 131, 133, 136-139, 141, 142, 144-147, 149, 153, 154, 161, 164, 166, 177, 179, 180, 183, 185, 193, 195-198, 201, 203-205, 207-213, 219, 220, 227, 229, 230, 234, 236, 238, 239, 244

Fiberglass, 9, 10, 16, 17, 19, 21-25, 29, 31, 33, 38, 53, 54, 59, 61, 65, 75-80, 82, 83, 89, 98, 105, 110, 125, 127, 131-135, 138-141, 146, 147, 149, 153, 154, 164, 166, 176, 177, 179-181, 183, 186-191, 193, 195, 197-199, 201, 204, 205, 208, 210-212, 214, 216, 219, 220, 227-229, 235, 236, 243, 250, 251, 261, 266, 270

Fibers, miscellaneous
 alumina, 34, 35, 105-107
 ceramic, 35
 silicon carbide, 35, 36, 105-107
 other (steel, nylon), 54

Filament winding, 9, 17, 33, 39-51, 69, 77, 79, 80, 95, 200, 201, 220, 222, 228, 230, 233, 235-237, 243

Graphite Fibers (See Carbon Graphite Fibers)

Hand layup (layup, stacked up fabrication), 12, 38, 65, 77, 79, 80, 190, 195-197, 201, 205, 214, 218, 220, 235, 236, 251, 252, 256, 257, 261, 263, 268, 270

Injection molding, 20, 79, 80, 165, 176, 186

Laminating techniques, 139–142, 251, 252, 266, 268

Marine applications, 9, 24, 38, 56, 83, 233, 275

Polyester, unsaturated, resins, 7–11, 19, 21, 54, 61, 79–81, 94, 137, 145, 146, 154, 177, 195, 227, 229, 230, 236, 238, 239, 243, 251, 257, 266, 268
Polyimide resins, 13–21, 94, 127, 131, 142, 143, 145–148, 165, 166
Prepregs, 3, 4, 8, 14, 15, 17, 19, 20, 26, 27, 30, 33, 59, 60, 68, 135, 138–141, 150, 151, 190, 195–197, 200, 201, 214, 215
Pressure vessels, 7, 24, 33, 42, 44, 83, 234
Process industry applications, 12, 24, 38, 39, 41, 56, 69, 227–248, 278
Programmable controllers (*See* Automation)
Pulforming, (See also Pultrusion), 62–65, 69

Pultrusion, 8, 12, 39, 47, 49, 52–65, 69, 80, 244, 251, 252, 256, 261, 263

Resin transfer molding (RTM), 4, 8, 80, 176, 190, 251, 252

Sheet molding compound (SMC), 8, 12, 13, 24, 79, 80, 176–179, 183, 187–190
Sporting goods, 9, 24, 27, 31, 39, 47, 56, 69, 83
Spray technique, 9, 38, 80
Stacked up fabrication, (See also Hand layup), 17, 140

Tape layup, 65–70

Vacuum bag fabrication/curing, 20, 38, 197, 200, 205, 217, 218
Vinyl ester resins, 1, 9, 11–13, 229, 230, 236, 238, 239, 251, 266, 268, 270